Nuclear Energy Policy

Mark Holt
Specialist in Energy Policy

June 20, 2012

Congressional Research Service
7-5700
www.crs.gov
RL33558

Summary

Nuclear energy issues facing Congress include power plant safety and regulation, radioactive waste management, research and development priorities, federal incentives for new commercial reactors, nuclear weapons proliferation, and security against terrorist attacks.

The earthquake and resulting tsunami that severely damaged Japan's Fukushima Daiichi nuclear power plant on March 11, 2011, raised questions in Congress about the disaster's possible implications for nuclear safety regulation, U.S. nuclear energy expansion, and radioactive waste policy. The tsunami knocked out all electric power at the six-reactor plant, resulting in the overheating of several reactor cores, loss of cooling in spent fuel storage pools, major hydrogen explosions, and releases of radioactive material to the environment. The Nuclear Regulatory Commission (NRC) issued orders to U.S. nuclear plants March 12, 2012, to begin implementing safety improvements in response to Fukushima.

Significant incentives for new commercial reactors were included in the Energy Policy Act of 2005 (EPACT05, P.L. 109-58), such as tax credits and loan guarantees. Together with volatile fossil fuel prices and the possibility of greenhouse gas controls, the federal incentives for nuclear power helped spur renewed interest by utilities and other potential reactor developers. License applications for as many as 31 new reactors have been announced, and NRC issued licenses for four reactors at two plant sites in early 2012. However, falling natural gas prices and other circumstances have made it unlikely that many more of the proposed nuclear projects will move toward construction in the near term.

DOE's nuclear energy research and development program includes advanced reactors, fuel cycle technology and facilities, and infrastructure support. The Obama Administration's FY2013 funding request totals $770.4 million, which is $88.3 million (10.3%) below the enacted FY2012 funding level. DOE is requesting $65 million for FY2013 to provide technical support for licensing small modular light water reactors (LWRs), $2 million below the FY2012 funding level. The House-passed version of the FY2013 Energy and Water appropriations bill (H.R. 5325) increased nuclear R&D by $89.9 million from FY2012, while the Senate Appropriations Committee recommended a $20.1 million increase (S. 2465).

Disposal of highly radioactive waste has been one of the most controversial aspects of nuclear power. The Nuclear Waste Policy Act of 1982 (P.L. 97-425), as amended in 1987, required DOE to conduct a detailed physical characterization of Yucca Mountain in Nevada as a permanent underground repository for high-level waste. The Obama Administration decided to "terminate the Yucca Mountain program while developing nuclear waste disposal alternatives," according to the DOE FY2010 budget justification. Alternative waste management strategies were evaluated by the Blue Ribbon Commission on America's Nuclear Future, which issued its final report to the Secretary of Energy on January 26, 2012. The report recommended options for temporary storage, treatment, and permanent disposal of highly radioactive nuclear waste, along with an evaluation of nuclear waste technologies. It did not recommend specific sites for new nuclear waste facilities or evaluate the suitability of Yucca Mountain. No funding was provided in FY2012 or requested for FY2013 to continue NRC licensing of the Yucca Mountain repository, although the issue is currently the subject of a federal appeals court case. The House-passed FY2013 Energy and Water bill provided DOE with $25 million to resume Yucca Mountain licensing, along with $10 million for NRC. The Senate Appropriations Committee authorized a pilot program to develop one or more voluntary nuclear waste storage sites.

Contents

Tables

Contacts

Most Recent Developments

The Nuclear Regulatory Commission (NRC) on February 9, 2012, approved the first licenses to build new U.S. commercial nuclear reactors in more than three decades. The combined operating licenses (COLs) allow Southern Company to construct and operate two new Westinghouse AP1000 reactors at the Vogtle nuclear power plant in Georgia. On March 30, 2012, NRC approved COLs for two additional AP1000 reactors at the existing Summer nuclear plant in South Carolina. Each of the new reactors, scheduled for completion between 2016 and 2019, is expected to cost from $5 billion to $7 billion.

NRC on March 12, 2012, issued its first nuclear plant safety requirements based on lessons learned from the March 2011 Fukushima disaster in Japan. NRC ordered U.S. nuclear plant operators to begin implementing safety enhancements related to the loss of power caused by natural disasters, reactor containment venting, and monitoring the water levels of reactor spent fuel pools. The Fukushima nuclear plant was hit by an earthquake and tsunami that knocked out all electric power at the six-reactor plant, resulting in the overheating of the reactor cores in three of the units and a heightened overheating risk at several spent fuel storage pools at the site. The overheating of the reactor cores caused major hydrogen explosions and releases of radioactive material to the environment. Several House and Senate hearings have been held on the accident, and several bills on nuclear safety have been introduced in the 112th Congress. Proposed bills would delay all new nuclear licenses and permits until stronger safety standards were in place (H.R. 1242), expand evacuation planning around U.S. nuclear reactors (H.R. 1268), and initiate U.S. efforts to strengthen international nuclear safety agreements (S. 640, H.R. 1326).

The Obama Administration requested $770.4 million for nuclear energy research and development in its FY2013 budget, submitted to Congress February 12, 2012. Including advanced reactors, fuel cycle technology, infrastructure support, and safeguards and security, the total nuclear energy request is $88.3 million (10.3%) below the enacted FY2012 funding level. Funding for safeguards and security in FY2012 was provided under a separate appropriations account, Other Defense Activities, but it is included under the Nuclear Energy account in the FY2013 request. The largest proposed reductions for FY2013 are Reactor Concepts (35.9%), Radiological Facility Management (26.6%), and Nuclear Energy Enabling Technologies (12.5%). The House-passed FY2013 Energy and Water Development appropriations bill would increase nuclear R&D by $89.9 million from FY2012 (H.Rept. 112-462), while the Senate Appropriations Committee recommended a $20.1 million increase (S.Rept. 112-164),

Nuclear energy funding for FY2012 was included in the Consolidated Appropriations Act, 2012 (P.L. 112-74), approved by Congress December 17, 2011. The funding measure included $67 million to commercialize small modular reactors and $60 million for nuclear waste disposal research.

The Blue Ribbon Commission on America's Nuclear Future, established by the Obama Administration to recommend a new strategy for nuclear waste management, issued its final report to the Secretary of Energy on January 26, 2012.[1] President Obama has moved to terminate previous plans to open a national nuclear waste repository at Yucca Mountain, NV. In its final

---

[1] Blue Ribbon Commission on America's Nuclear Future, *Report to the Secretary of Energy*, January 2012, http://brc.gov/sites/default/files/documents/brc_finalreport_jan2012.pdf.

report, the Blue Ribbon Commission recommended a "consent-based" approach to siting nuclear waste facilities and that the roles of local, state, and tribal governments be negotiated for each potential site. The development of consolidated waste storage and disposal facilities should begin as soon as possible, the Commission urged. A new waste management organization should be established to develop the repository, along with associated transportation and storage systems, according to the Commission. The new organization should have "assured access" to the Nuclear Waste Fund, which holds fees collected from nuclear power plant operators to pay for waste disposal. Under existing law, the Nuclear Waste Fund cannot be drawn down without congressional appropriations. The House appropriations bill would provide DOE with $25 million for FY2013 to resume Yucca Mountain licensing, along with $10 million for NRC in a floor amendment (H.R. 5325). The Senate Appropriations Committee on April 26, 2012, authorized a pilot program to develop one or more voluntary nuclear waste storage sites (S. 2465).

President Obama's State of the Union Address on January 25, 2011, called for nuclear power to be included in a national goal of generating 80% of U.S. electricity "from clean energy sources" by 2035. Along with nuclear power and renewable energy, "clean energy" would include "efficient" natural gas plants and clean coal technologies, to the extent that they reduced carbon emissions compared with conventional coal-fired plants. The President's proposed Clean Energy Standard could provide a significant boost to U.S. nuclear power expansion, particularly in areas of the country with relatively limited renewable energy resources. Senator Bingaman, Chairman of the Senate Energy and Natural Resources Committee, introduced legislation to establish a national clean energy standard March 1, 2012 (S. 2146).

Nuclear Power Status and Outlook

After nearly 30 years in which no new orders had been placed for nuclear power plants in the United States, a series of license applications that began in 2007 prompted widespread speculation about a U.S. "nuclear renaissance." The renewed interest in nuclear power largely resulted from the improved performance of existing reactors, federal incentives in the Energy Policy Act of 2005 (P.L. 109-58), the possibility of carbon dioxide controls that could increase costs at fossil fuel plants, and volatile prices for natural gas—the favored fuel for new power plants for the past two decades.

Four of the proposed new U.S. reactors received licenses from the Nuclear Regulatory Commission (NRC) in early 2012. NRC approved combined construction permit and operating licenses (COLs) for Southern Company to build and operate two new Westinghouse AP1000 reactors at the Vogtle nuclear power plant in Georgia on February 9, 2012. On March 30, 2012, NRC approved COLs for two additional AP1000 reactors at the existing Summer nuclear plant in South Carolina. Substantial site preparation and infrastructure work has already taken place at both sites, and the owners of both projects announced plans to move to full construction after receiving their COLs.[2]

---

[2] Southern Company, "Southern Company Subsidiary Receives Historic License Approval for New Vogtle Units, Full Construction Set to Begin," February 9, 2012, http://www.southerncompany.com/news/iframe_pressroom.aspx; SCANA, "NRC Approves COLs for SCE&G, Santee Cooper Nuclear Units," March 30, 2012, http://www.scana.com/en/investor-relations/news-releases/nrc-approves-cols-for-sceg-santee-cooper-nuclear-units.htm.

However, the future of all other proposed new U.S. reactors is uncertain. High construction cost estimates—a major reason for earlier reactor cancellations—continue to undermine nuclear power economics. A more recent obstacle to nuclear power growth has been the development of vast reserves of domestic natural gas from previously uneconomic shale formations, which has held gas prices low and reduced concern about future price spikes. Moreover, uncertainty over U.S. controls on carbon emissions may be further increasing caution by utility companies about future nuclear projects.

The March 11, 2011, earthquake and tsunami that severely damaged Japan's Fukushima Daiichi nuclear power plant could also affect plans for new U.S. reactors, although U.S. nuclear power growth was already expected to be modest in the near term. Following the Fukushima accident, preconstruction work was suspended on two planned reactors at the South Texas Project. Tokyo Electric Power Company (TEPCO), which owns the Fukushima plant, had planned to invest in the South Texas Project expansion, but TEPCO's financial condition plunged after the accident. New U.S. safety requirements resulting from the Fukushima disaster could raise investor concerns about higher costs. On the other hand, after the accident the Obama Administration reiterated its support for nuclear power expansion as part of its clean energy policy.[3]

The recent applications for new power reactors in the United States followed a long period of declining nuclear generation growth rates. Until the COLs were issued for the Vogtle and Summer projects, no nuclear power plants had been ordered in the United States since 1978, and more than 100 reactors had been canceled, including all ordered after 1973. The most recent U.S. nuclear unit to be completed was the Tennessee Valley Authority's (TVA's) Watts Bar 1 reactor, ordered in 1970 and licensed to operate in 1996. But largely because of better operation and capacity expansion at existing reactors, annual U.S. nuclear generation has risen by about 20% since the startup of Watts Bar 1.[4]

The U.S. nuclear power industry currently comprises 104 licensed reactors at 65 plant sites in 31 states and generates about 20% of the nation's electricity.[5] TVA's board of directors voted August 1, 2007, to resume construction on Watts Bar 2, which had been suspended in 1985; the renewed construction project was to cost about $2.5 billion and be completed in 2013. However, TVA announced on April 5, 2012, that completing Watts Bar 2 would cost up to $2 billion more than expected and take until 2015.[6] At TVA's request, NRC in March 2009 reinstated the construction authorization for the two-unit Bellefonte (AL) nuclear plant, which had been deferred in 1988 and canceled in 2006.[7] The TVA board voted on August 18, 2011, to complete construction of Bellefonte 1 after the Watts Bar 2 project is finished. Completing Bellefonte 1 was projected at that time to cost $4.9 billion, with operation to begin by 2020.[8]

---

[3] Oral Testimony of Energy Secretary Steven Chu at the House Energy and Commerce Committee – As Prepared for Delivery, March 16, 2011, http://www.energy.gov/news/10178.htm.

[4] Energy Information Administration, *Electric Power Monthly*, Net Generation by Energy Source, April 2011, http://www.eia.gov/cneaf/electricity/epm/epm_sum.html.

[5] U.S. Nuclear Regulatory Commission, *Information Digest 2008-2009*, NUREG-1350, Vol. 20, August 2008, p. 32, http://www.nrc.gov/reading-rm/doc-collections/nuregs/staff/sr1350/v20/sr1350v20.pdf.

[6] Mary Powers, "Credit Agencies See Watts Bar-2 Cost Impact," *Nucleonics Week*, April 12, 2012, p. 1.

[7] Nuclear Regulatory Commission, "In the Matter of Tennessee Valley Authority (Bellefonte Nuclear Plant Units 1 and 2)," 74 *Federal Register* 10969, March 13, 2009.

[8] Tennessee Valley Authority, "TVA Board Implements Vision," press release, August 18, 2011, http://www.tva.com/news/releases/julsep11/board_meeting/index.htm.

Annual electricity production from U.S. nuclear power plants is much greater than that from oil and hydropower and other renewable energy sources. Nuclear generation has been overtaken by natural gas in recent years, and it remains well behind coal, which accounts for about 45% of U.S. electricity generation.[9] Nuclear plants generated more than half the electricity in four states in 2011—Connecticut, New Jersey, South Carolina, and Vermont.[10] The 790 billion net kilowatt-hours of nuclear electricity generated in the United States during 2011[11] was about the same as the nation's entire electrical output in the early 1960s, when the oldest of today's operating U.S. commercial reactors were ordered.[12]

Reasons for the 30-year halt in U.S. nuclear plant orders included high capital costs, public concern about nuclear safety and waste disposal, and regulatory compliance issues.

High construction costs may pose the most serious obstacle to nuclear power expansion. Construction costs for reactors completed since the mid-1980s ranged from $2 to $6 billion, averaging more than $3,900 per kilowatt of electric generating capacity (in 2011 dollars), far higher than commercial fossil fuel technologies. The nuclear industry predicts that new plant designs could be built for less than that if many identical plants were built in a series, but current estimates for new reactors show little if any reduction in cost.[13]

In contrast, average U.S. nuclear plant operating costs per kilowatt-hour dropped substantially since 1990, and expensive downtime has been steadily reduced. Licensed U.S. commercial reactors generated electricity at an average of 89% of their total capacity in 2011, according to the Energy Information Administration (EIA).[14]

Seventy-three commercial reactors have received 20-year license extensions from the Nuclear Regulatory Commission (NRC), giving them up to a total of 60 years of operation. License extensions for 13 additional reactors are currently under review, and more are anticipated, according to NRC.[15] The FY2012 Consolidated Appropriations Act (P.L. 112-74) provided $25 million for DOE to study further reactor life extension to 80 years, and DOE requested $21.7 million for that program in FY2013.

Existing nuclear power plants appear to hold a strong position in electricity wholesale markets. In most cases, nuclear utilities have received favorable regulatory treatment of past construction costs, and average existing nuclear plant operating costs are estimated to be competitive with

---

[9] Energy Information Administration, *Electric Power Monthly*, Net Generation by Energy Source, February 2012, http://www.eia.gov/cneaf/electricity/epm/epm_sum.html. Net generation excludes electricity used for power plant operation.

[10] Nuclear Regulatory Commission, *Information Digest, 2011–2012*, NUREG-1350, Volume 23, http://www.nrc.gov/reading-rm/doc-collections/nuregs/staff/sr1350/v23/sr1350v23-sec-2.pdf.

[11] Ibid.

[12] All of today's 104 operating U.S. commercial reactors were ordered from 1963 through 1973; see "Historical Profile of U.S. Nuclear Power Development," U.S. Council for Energy Awareness, 1992.

[13] For a comparison of generating costs, see CRS Report RL34746, *Power Plants Characteristics and Costs*, by Stan Mark Kaplan.

[14] Energy Information Administration, "U.S. Nuclear Generation and Generating Capacity," http://www.eia.gov/cneaf/nuclear/page/nuc_generation/gensum.html.

[15] Nuclear Regulatory Commission, *Fact Sheet on Reactor License Renewal*, August 8, 2011, http://www.nrc.gov/reading-rm/doc-collections/fact-sheets/fs-reactor-license-renewal.html.

those of fossil fuel technologies.[16] Although eight U.S. nuclear reactors were permanently shut down during the 1990s, none has been closed since 1998.

Possible New Reactors

Electric utilities and other firms have announced plans to apply for COLs for more than 30 reactors (see **Table 1**).[17] (For a discussion of COLs, see the "Licensing and Regulation" section below.)

As noted above, construction is currently underway on four of the proposed new reactors, at the Vogtle and Summer sites. COLs are being actively pursued for about 16 additional reactors (shown in **Table 1**), whose owners have not committed to actual construction but are keeping the option available if conditions are more favorable in the future. The experience of the first few reactors to be constructed is likely to be crucial in determining whether a wave of subsequent units will move forward as the nuclear industry envisions.

The two new Vogtle reactors are scheduled to go on line in 2016 and 2017,[18] while the Summer units are planned for 2016 and 2019.[19] EIA estimates that construction costs of new nuclear power plants will average $5,335 per kilowatt of capacity, or about $6.1 billion for an AP1000 unit, not including interest costs.[20] The two Summer units are expected to cost about $11.6 billion in 2012 dollars, according to regulatory filings,[21] while the two Vogtle units are projected by their primary owner to cost a total of $13.35 billion.[22]

Progress Energy's Levy County project, with two AP1000 units, is scheduled by NRC to receive a final decision on its COLs in early 2013. As shown in **Table 1**, the remaining 9 projects that are actively seeking COLs, with a total of 14 proposed reactors, do not have firm licensing schedules from NRC. Several of those projects would use designs that also do not have firm NRC review schedules. As a result, these reactors appear unlikely to be completed before the early 2020s. This group includes the planned units 3 and 4 at the South Texas Project, where preconstruction work was suspended after the Fukushima Daiichi accident, as noted above. The joint venture developing the new South Texas Plant reactors, Nuclear Innovation North America (NINA), will focus solely on the COL and a DOE loan guarantee.[23] Several of these proposed nuclear projects

---

[16] Energy Information Administration, *Nuclear Power 12 percent of America's Generating Capacity, 20 percent of the Electricity*, July 17, 2003, at http://www.eia.doe.gov/cneaf/nuclear/page/analysis/nuclearpower.html.

[17] Nuclear Regulatory Commission, New Reactors, http://www.nrc.gov/reactors/new-reactors.html.

[18] Southern Company, "Smart Power," http://www.southerncompany.com/smart_energy/smart_power_vogtle-kemper.html.

[19] Shaw Group, "V.C. Summer Nuclear Station, Units 2 and 3," http://www.shawgrp.com/projects/nuclear/vcsummer.

[20] Energy Information Administration, "Updated Capital Cost Estimates for Electricity Generation Plants," November 2010, http://www.eia.gov/oiaf/beck_plantcosts/index.html.

[21] South Carolina Electric & Gas Company, "Petitions for Updates and Revisions to the Capital Cost Schedule and the Construction Schedule," before the Public Service Commission of South Carolina, February 29, 2012, http://www.scana.com/NR/rdonlyres/35AAED95-5226-416A-8DC2-0743BC93B911/0/2012PetitiontoUpdateCostSchedules.pdf. Total cost based on SCE&G ownership of 55%.

[22] Southern Company, "Smart Power," http://www.southerncompany.com/smart_energy/smart_power_vogtle-kemper.html. Total cost based on Southern Company's 45.7% ownership.

[23] NRG Energy, "NRG Energy, Inc. Provides Greater Clarity on the South Texas Nuclear Development Project," press release, April 19, 2011, http://phx.corporate-ir.net/External.File?item=UGFyZW50SUQ9OTAwMzB8Q2hpbGRJRD0tMXxUeXBlPTM=&t=1.

---

may require additional partners in order to proceed to construction, according to recent company announcements.[24]

Several other COL applications have been suspended, withdrawn, or shifted to early site permits (ESPs) only. Entergy suspended further license review of its planned GE ESBWR reactors at River Bend, LA, and Grand Gulf, MS, although it still has a previously issued ESP for Grand Gulf. AmerenUE suspended review of a COL for its proposed new Callaway unit in Missouri, and Exelon withdrew its COL application for a proposed two-unit plant in Victoria County, TX, but both are now seeking early site permits instead, laying the groundwork for possible future licensing.

TVA decided to defer consideration of its COL application for two new Westinghouse AP1000 reactors at its Bellefonte plant in Alabama in favor of completing the first of two unfinished Babcock & Wilcox reactors at the site. TVA had submitted a COL application for the Bellefonte AP1000s in October 2007 as part of the NuStart consortium.[25]

Constellation Energy announced October 9, 2010, that it was abandoning negotiations with DOE for a loan guarantee for the planned Calvert Cliffs 3 reactor, which Constellation had been developing as part of its UniStar joint venture with the French national utility EDF.[26] Constellation sold its share of UniStar to EDF so that EDF could seek another U.S. partner to continue the Calvert Cliffs project.[27] (For more discussion of Constellation's decision, see the "Loan Guarantees" section below.)

NRC anticipates that several more COL and other license applications will be submitted in the next two years. This includes a TVA plan to submit construction permit applications for six small modular reactors (SMRs) of about 160 megawatts each at its Clinch River, TN, site.

---

[24] Jeff Beattie, "Southeast Utilities Seek Partners to Hedge Nuclear Bets," *Energy Daily*, October 5, 2010, p. 1.

[25] Tennessee Valley Authority, "Single Nuclear Unit at the Bellefonte Plant Site," fact sheet, http://www.tva.gov/environment/reports/blnp/index.htm.

[26] Constellation Energy, "Constellation Energy Releases Statement Regarding U.S. Department of Energy Loan Guarantee," press release, October 9, 2010, http://ir.constellation.com/releasedetail.cfm?ReleaseID=516614.

[27] Letter from Michael J. Wallace, Vice Chairman and Chief Operating Officer, Constellation Energy, to Thomas Piquemal, Group Executive Vice President, Finance, EDF, October 15, 2010, http://files.shareholder com/downloads/CEG/1036755503x0x410084/e27369a0-ce85-432f-bfad-e17ddce4f8f2/101510_-_EDF_letter.pdf; Unistar, "EDF and Constellation Energy Announce Comprehensive Agreement," press release, October 27, 2010, http://press.edf.com/press-releases/all-press-releases/2010/edf-and-constellation-energy-announce-comprehensive-agreement-82018.html&return=42873.

Table 1. Announced Nuclear Plant License Applications

Announced Applicant	Site	Reactor Type	Units	Status
COL issued				
Southern	Vogtle (GA)	Westinghouse AP1000	2	COL app ication submitted 3/13/08; engineering, procurement, and construction (EPC) contract signed 4/8/08; ESP and limited construction approved 8/26/09; conditional DOE loan guarantee announced 2/16/10; NRC hearing held 9/27-28/11; COL approved February 9, 2012
SCE&G	Summer (SC)	Westinghouse AP1000	2	COL submitted 3/31/08; EPC contract signed 5/27/08; COL approved March 30, 2012
COL scheduled for completion				
Progress Energy	Levy County (FL)	Westinghouse AP1000	2	COL submitted 7/30/08; scheduled for completion in 2013
COL schedule under revision				
DTE Energy	Fermi (MI)	GE ESBWR	1	COL submitted 9/18/08
FPL	Turkey Point (FL)	Westinghouse AP1000	2	COL submitted 6/30/09; preconstruction work being conducted
Luminant Power (formerly TXU)	Comanche Peak (TX)	Mitsubishi US-APWR	2	COL submitted 9/19/08
Duke Energy	Wil iam States Lee (SC)	Westinghouse AP1000	2	COL submitted 12/13/07
Nuclear Innovation North America	South Texas Project	Toshiba ABWR	2	COL submitted 9/20/07; EPC contract signed with Toshiba 2/12/09; NRG Energy halted further investment 4/19/11
PPL	Bell Bend (PA)	Areva EPR	1	COL submitted 10/10/08
Progress Energy	Harris (NC)	Westinghouse AP1000	2	COL submitted 2/19/08; EPC contract signed 1/5/09
UniStar	Calvert C iffs (MD)	Areva EPR	1	COL submitted 7/13/07 (Part 1), 3/13/08 (Part 2); Constellation withdrew from project 10/8/10
Dominion	North Anna	Mitsubishi US-APWR	1	COL submitted 11/27/07; ESP approved 11/20/07; reactor selection announced 5/7/10

Announced Applicant	Site	Reactor Type	Units	Status
Licensing suspended				
Entergy	Grand Gu f (MS)	Not specified	I	COL submitted 2/27/08; licensing suspended 1/9/09; ESP approved 3/27/07
Exelon	Victoria County (TX)	Not specified	2	COL app ication withdrawn and ESP application submitted 3/25/10
AmerenUE	Calloway (MO)	Areva EPR	I	COL submitted 7/24/08; license review suspended 6/23/09; ESP expected 2012
Entergy	River Bend (LA)	Not specified	I	COL submitted 9/25/08; licensing suspended 1/9/09
TVA	Bellefonte	Westinghouse AP1000	2	COL submitted 10/30/07; licensing deferred 9/29/10
Unistar	Nine Mile Point (NY)	Areva EPR	I	COL submitted 9/30/08; licensing suspended 12/1/09
Anticipated license applications				
Blue Castle	Utah	Not specified	I	ESP application expected in 2012
TVA	C inch River (TN)	mPower small modular reactor	6	Construction permit application expected in 2014; operating license application expected in 2017
AmerenUE	Missouri	Westing. SMR	I	COL app ication expected in 2012
Unnamed	Unspecified	Unspecified	I	COL app ication expected in 2013
Southern	Unspecified	Unspecified	I	COL app ication expected in 2013
Total units announced			**38**	
Total currently active COLs			**20**	

Sources: NRC, *Nucleonics Week*, *Nuclear News*, Nuclear Energy Institute, company news releases.

Note: Applications are for COLs unless otherwise specified.

Nuclear Power Plant Safety and Regulation

Safety

Worldwide concern about nuclear power plant safety rose sharply after the Fukushima accident, which is generally considered to be much worse than the March 1979 Three Mile Island accident in Pennsylvania but not as severe as the April 1986 Chernobyl disaster in the former Soviet Union. Based on dose rates reported by Japanese authorities, the Natural Resources Defense Council (NRDC) estimated that the Fukushima accident subjected the population to a total radiation dose of 148,000 person-rem through April 5. In comparison, the total dose from Three Mile Island was estimated at 2,000 person-rem, while Chernobyl was estimated at 25.5 million person-rem.[28] The Fukushima disaster resulted in similar levels of radioactive contamination per

---

[28] Matthew McKinzie and Thomas B. Cochran, Natural Resources Defense Council, "The Collective Effective Dose (continued...)

square meter to that of Chernobyl, but the Fukushima contamination was much less widespread and affected a smaller number of people.[29] (For more background on the Fukushima accident, see CRS Report R41694, *Fukushima Nuclear Disaster*, by Mark Holt, Richard J. Campbell, and Mary Beth Nikitin.)

The Fukushima accident has raised particular policy questions for the United States because, unlike Chernobyl, the Fukushima reactors are similar to common U.S. designs. Although the Fukushima accident resulted from a huge tsunami that incapacitated the power plant's emergency diesel generators, the accident dramatically illustrated the potential consequences of any natural catastrophe or other situation that could cause an extended "station blackout" – the loss of alternating current (AC) power. Safety issues related to station blackout include standards for backup batteries, which now are required to provide power for 4-8 hours, and additional measures that may be required to assure backup power. The Institute of Nuclear Power Operations (INPO) released a detailed description of the Fukushima accident in November 2011.[30]

Safety concerns at U.S. reactors were also raised by hydrogen explosions at four of the Fukushima reactors—resulting from a high-temperature reaction between steam and nuclear fuel cladding—and the loss of cooling at the Japanese plant's spent fuel storage pools. Other safety issues that have been raised in the wake of Fukushima include the vulnerability of U.S. nuclear plants to earthquakes, floods, and other natural disasters, the availability of iodine pills to prevent absorption of radioactive iodine released during nuclear accidents, and the adequacy of nuclear accident emergency planning.

In response to such concerns, NRC on March 23, 2011, established a task force "made up of current senior managers and former NRC experts" to "conduct both short- and long-term analysis of the lessons that can be learned from the situation in Japan."[31] The Near-Term Task Force issued its report July 12, 2011, making recommendations ranging from specific safety improvements to broad changes in NRC's overall regulatory approach.[32] NRC staff subsequently identified several of those actions that "can and should be initiated without delay."[33] The NRC Commissioners largely agreed with the recommendations on October 18, 2011, and instructed the agency's staff

---

(...continued)

Resulting from Radiation Emitted During the First Weeks of the Fukushima Daiichi Nuclear Accident," April 10, 2011, http://docs.nrdc.org/nuclear/files/nuc_11041301a.pdf. A person-rem is the equivalent of one person receiving a radiation dose of one rem. For background on radiation doses, see CRS Report R41728, *The Japanese Nuclear Incident Technical Aspects*, by Jonathan Medalia.

[29] French Institut de Radioprotection et de Surete Nucleaire (IRSN), Assessment on the 66th Day of Projected External Doses for Populations Living in the North-West Fallout Zone of the Fukushima Nuclear Accident, Report DRPH/2011-10, p. 27, http://www.irsn.fr/EN/news/Documents/IRSN-Fukushima-Report-DRPH-23052011.pdf.

[30] Institute of Nuclear Power Operations, *Special Report on the Nuclear Accident at the Fukushima Daiichi Nuclear Power Station*, INPO 11-005, November 2011, available from the Nuclear Energy Institute at http://www.nei.org/resourcesandstats/documentlibrary/safetyandsecurity/reports/special-report-on-the-nuclear-accident-at-the-fukushima-daiichi-nuclear-power-station.

[31] Nuclear Regulatory Commission, "Nuclear Regulatory Commission Directs Staff on Continuing Agency Response to Japan Events; Adjust Commission Schedule," press release, March 23, 2011, http://pbadupws.nrc.gov/docs/ML1108/ML110821123.pdf.

[32] Near-Term Task Force Review of Insights from the Fukushima Dai-ichi Accident, *Recommendations for Enhancing Reactor Safety in the 21st Century*, Nuclear Regulatory Commission, Washington, DC, July 12, 2011, http://pbadupws.nrc.gov/docs/ML1118/ML111861807.pdf.

[33] NRC, "Recommended Actions to Be Taken Without Delay from the Near-Term Task Force Report," SECY-11-0124, September 9, 2011.

---

to "strive to complete and implement the lessons learned from the Fukushima accident within five years—by 2016."[34] Tier 1 regulatory actions, which are to get underway immediately, include:

- *Seismic and flood hazard reevaluations and walkdowns.* Nuclear plant operators will be required to evaluate the implications of updated seismic and flooding models, including all potential flooding sources. Plant operators will be required to identify and verify the adequacy of flood and seismic protection features at their sites.

- *Station blackout regulatory actions.* NRC will issue an advance notice of proposed rulemaking (ANPR) with the goal of requiring that nuclear power plants be able to cope with the total loss of AC power (station blackout) for at least eight hours. The eight hour period is intended to give plant personnel enough time to restore AC power or, if that is not possible, to take actions to extend the plant's ability to cope with the loss of AC power to at least 72 hours. The eight-hour coping time would rely only on permanently installed equipment, while the 72-hour coping time could rely on off-site, portable equipment. Enough equipment and personnel would be required to protect all affected reactors at a multi-unit plant. While new regulations are being prepared, NRC is to order plant operators to protect emergency equipment from damage from external events and ensure that enough equipment is available to protect all reactors at a plant site.

- *Reliable hardened vents for Mark I containments.* NRC will order nuclear plants to install vents for the containments in Mark I reactors (the type at Fukushima). The vents would be designed to reduce containment pressure while preventing hydrogen in the containment from leaking into the reactor building, as occurred at Fukushima.

- *Spent fuel pool instrumentation.* NRC will order nuclear plants to install safety instrumentation to monitor spent fuel pool conditions, such as water level, temperature, and radiation levels, from the plant control room.

- *Strengthening and integrating accident procedures and guidelines.* NRC will order nuclear plants to modify emergency operating procedures to integrate severe accident management guidelines and extensive damage mitigation guidelines. The modifications would have to specify clear command-and-control strategies and establish training qualifications for emergency decisionmakers.

- *Emergency preparedness regulatory actions.* Pending a rulemaking, NRC will order nuclear plants to ensure adequate emergency preparedness training for multi-reactor station blackouts and other emergencies.

The NRC staff slightly modified its proposals for top priority actions and divided the remaining Task Force proposals into two lower tiers, which were determined to require further assessment and potentially long-term study. Included in the lower-tier actions were requirements for emergency water supply systems for spent fuel pools, secure power for emergency communications and data systems, confirmation of seismic and flooding hazards, and modifications to NRC's regulatory process.[35]

---

[34] NRC, "Staff Requirements – SECY-11-0124 – Recommended Actions to Be Taken Without Delay from the Near-Term Task Force Report," October 18, 2011, http://pbadupws.nrc.gov/docs/ML1126/ML11269A204.pdf.

[35] R.W. Borchardt, NRC Executive Director for Operations, "Prioritization of Recommended Actions to Be Taken in (continued...)

On March 12, 2012, NRC issued its first nuclear plant safety requirements based on the lessons learned from Fukushima. NRC ordered U.S. nuclear plant operators to begin implementing safety enhancements related to the loss of power caused by natural disasters, reactor containment venting, and monitoring the water levels of reactor spent fuel pools. Nuclear plant operators were required to begin implementing the requirements immediately and come into full compliance no later than the end of 2016.[36] NRC also issued an advance notice of proposed rulemaking for new regulatory actions on station blackout March 20, 2012.[37]

Legislation introduced after the Fukushima accident includes the Nuclear Power Plant Safety Act of 2011 (H.R. 1242), introduced by Representative Markey on March 29, 2011. It would require NRC to revise its regulations within 18 months to ensure that nuclear plants could handle major disruptive events, a loss of off-site power for 14 days, and the loss of diesel generators for 72 hours. Spent fuel would have to be moved from pool to dry-cask storage within a year after it had cooled sufficiently, and emergency planning would have to include multiple concurrent disasters. NRC could not issue new licenses or permits until the revised regulations were in place.

Emergency Planning

Following the Three Mile Island accident, which revealed severe weaknesses in preparations for nuclear plant emergencies, Congress mandated that emergency plans be prepared for all licensed power reactors (P.L. 96-295, Sec. 109). NRC was required to develop standards for emergency plans and review the adequacy of each plant-specific plan in consultation with the Federal Emergency Management Agency (FEMA).

NRC's emergency planning requirements focus on a "plume exposure pathway emergency planning zone (EPZ)," encompassing an area within about 10 miles of each nuclear plant. Within the 10-mile EPZ, a range of responses must be developed to protect the public from radioactive releases, including evacuation, sheltering, and the distribution of non-radioactive iodine (as discussed above). The regulations also require a 50-mile "ingestion pathway EPZ," in which actions are developed to protect food supplies.[38] Nuclear plants are required to conduct emergency preparedness exercises every two years. The exercises, which are evaluated by FEMA and NRC, may include local, state, and federal responders and may involve both the plume and ingestion EPZs.[39]

The size of the plume exposure EPZ has long been a subject of controversy, particularly after the 9/11 terrorist attacks on the United States, in which nuclear plants were believed to have been a potential target. Attention to the issue was renewed by the Fukushima accident, in which some of the highest radiation dose rates have been measured beyond 10 miles from the plant.[40]

(...continued)

Response to Fukushima Lessons Learned," SECY-11-0137, October 3, 2011.

[36] Nuclear Regulatory Commission, "Actions in Response to the Japan Nuclear Accident," May 1, 2012, http://www.nrc.gov/reactors/operating/ops-experience/japan-info.html.

[37] Nuclear Regulatory Commission, "Station Blackout," Advance notice of proposed rulemaking, *Federal Register*, March 20, 2012, p. 16175, http://www.gpo.gov/fdsys/pkg/FR-2012-03-20/pdf/2012-6665.pdf.

[38] 10 CFR 50.47, Emergency Plans.

[39] Nuclear Regulatory Commission, "Emergency Preparedness & Response," website, http://www.nrc.gov/about-nrc/emerg-preparedness.html.

[40] Japanese Ministry of Education, Culture, Sports, Science, and Technology (MEXT), "Readings of Integrated Dose at (continued...)

Controversy over the issue intensified after NRC recommended on March 16, 2011, the evacuation of U.S. citizens within 50 miles of the Fukushima plant. The NRC recommendation was based on computer models that, using meteorological data and estimates of plant conditions, found that potential radiation doses 50 miles from the plant could exceed U.S. protective action guidelines.[41] Legislation introduced by Representative Lowey (H.R. 1268) would require evacuation planning within 50 miles of U.S. nuclear power plants.

In response to the 9/11 terrorist attacks, NRC modified its nuclear plant emergency planning requirements and began a comprehensive review of emergency planning regulations and guidance. The NRC staff sent a proposed final rule based on that review to the NRC Commissioners for approval on April 8, 2011, and the rule took effect December 23, 2011. [42] Among the changes included in the rule are new requirements for periodic updates of EPZ evacuation time estimates, mandatory backups for public alert systems, and protection of emergency responders during terrorist attacks. The new emergency planning regulations were prepared before the Fukushima accident, but the NRC staff recommended approval of the changes without waiting for further changes that might result from the lessons of the Japanese accident. Emergency planning changes resulting from Fukushima should be implemented later, the staff recommended.[43]

Domestic Reactor Safety Experience

Nuclear power safety has been a longstanding issue in the United States. Safety-related shortcomings have been identified in the construction quality of some plants, plant operation and maintenance, equipment reliability, emergency planning, and other areas. In one serious case, it was discovered in March 2002 that leaking boric acid had eaten a large cavity in the top of the reactor vessel in Ohio's Davis-Besse nuclear plant. The corrosion left only the vessel's quarter-inch-thick stainless steel inner liner to prevent a potentially catastrophic loss of reactor cooling water. Davis-Besse remained closed for repairs and other safety improvements until NRC allowed the reactor to restart in March 2004.

NRC's oversight of the nuclear industry is a subject of contention as well; nuclear utilities often complain that they are subject to overly rigorous and inflexible regulation, but nuclear critics charge that NRC frequently relaxes safety standards when compliance may prove difficult or costly to the industry.

In terms of public health consequences, the safety record of the U.S. nuclear power industry in comparison with other major commercial energy technologies has been excellent. During more than 3,500 reactor-years of operation in the United States,[44] the only incident at a commercial

---

(...continued)

Monitoring Post out of 20 Km Zone of Fukushima Dai-ichi NPP," data series, http://www.mext.go.jp/english/incident/ 1304275.htm.

[41] Nuclear Regulatory Commission, "NRC Provides Protective Action Recommendations Based on U.S. Guidelines," press release, March 16, 2011, http://pbadupws.nrc.gov/docs/ML1108/ML110800133.pdf.

[42] Nuclear Regulatory Commission, "Enhancements to Emergency Planning Regulations," Final rule, *Federal Register*, November 23, 2011, p. 72560.

[43] Nuclear Regulatory Commission, "Final Rule: Enhancements to Emergency Preparedness Regulations," SECY-11-0053, April 8, 2011, http://www.nrc.gov/reading-rm/doc-collections/commission/secys/2011/2011-0053scy.pdf.

[44] Nuclear Energy Institute, "Myths and Facts About Nuclear Energy," January 2012, p. 12, .http://www.nei.org/ (continued...)

nuclear power plant that might lead to any deaths or injuries to the public has been the Three Mile Island accident, in which more than half the reactor core melted.[45] A study of 32,000 people living within five miles of the reactor when the accident occurred found no significant increase in cancer rates through 1998, although the authors noted that some potential health effects "cannot be definitively excluded."[46]

The relatively small amounts of radioactivity released by nuclear plants during normal operation are not generally believed to pose significant hazards, although some groups contend that routine emissions are unacceptably risky. There is substantial scientific uncertainty about the level of risk posed by low levels of radiation exposure; as with many carcinogens and other hazardous substances, health effects can be clearly measured only at relatively high exposure levels. In the case of radiation, the assumed risk of low-level exposure has been extrapolated mostly from health effects documented among persons exposed to high levels of radiation, particularly Japanese survivors of nuclear bombing in World War II, medical patients, and nuclear industry workers.[47]

NRC announced April 7, 2010, that it had asked the National Academy of Sciences (NAS) to "perform a state-of-the-art study on cancer risk for populations surrounding nuclear power facilities." Unlike in previous studies, NAS is to examine cancer diagnosis rates, rather than cancer deaths, potentially increasing the amount of data. The new study would also use geographic units smaller than counties to determine how far members of the study group are located from reactors, to more clearly determine whether there is a correlation between cancer cases and distance from reactors.[48]

NRC's 1986 Safety Goal Policy Statement declared that nuclear power plants should not increase the risk of accidental or cancer deaths among the nearby population by more than 0.1%.[49] Later NRC guidance established a "subsidiary benchmark" for the probability of accidental core damage (fuel melting): Core damage frequency should average no more than one in 10,000 per reactor per year.[50] In addition, NRC set a benchmark that reactor containments should be successful at least 90% of the time in preventing major radioactive releases during a core-damage accident. Therefore, the benchmark probability of a major release from containment failure

---

(...continued)

resourcesandstats/documentlibrary/reliableandaffordablecnergy/factsheet/myths—facts-about-nuclear-energy-january-2012.

[45] Nuclear Regulatory Commission, "Backgrounder on the Three Mile Island Accident," March 15, 2011, http://www.nrc.gov/reading-rm/doc-collections/fact-sheets/3mile-isle.html.

[46] Evelyn O. Talbott et al., "Long Term Follow-Up of the Residents of the Three Mile Island Accident Area: 1979-1998," *Environmental Health Perspectives*, published online October 30, 2002, at http://ehp.niehs.nih.gov/docs/2003/5662/abstract.html.

[47] National Research Council, Committee to Assess the Health Risks from Exposure to Low Levels of Ionizing Radiation, *Beir VII Health Risks from Exposure to Low Levels of Ionizing Radiation, Report in Brief*, http://dels-old.nas.edu/dels/rpt_briefs/beir_vii_final.pdf.

[48] Nuclear Regulatory Commission, "NRC Asks National Academy of Sciences to Study Cancer Risk in Populations Living Near Nuclear Power Facilities," press release, April 7, 2010, http://www.nrc.gov/reading-rm/doc-collections/news/2010/10-060.html.

[49] NRC, "Safety Goals for the Operations of Nuclear Power Plants," policy statement, *Federal Register*, August 21, 1986, p. 30028, http://www.nrc.gov/reading-rm/doc-collections/commission/policy/51fr30028.pdf.

[50] NRC Staff Requirements Memorandum on SECY-89-102, "Implementation of the Safety Goals," Memorandum to James M. Taylor fro Samuel J. Chilk, June 15, 1990, http://pbadupws.nrc.gov/docs/ML0037/ML003707881.pdf.

---

during a core melt accident would average less than one in 100,000 per reactor per year.[51] (For the current U.S. fleet of about 100 reactors, that rate would yield an average of one core-damage accident every 100 years and a major release every 1,000 years.) On the other hand, some groups challenge the complex calculations that go into predicting such accident frequencies, contending that accidents with serious public health consequences may be more frequent.[52]

Reactor Safety in the Former Soviet Bloc

The Chernobyl accident was by far the worst nuclear power plant accident to have occurred anywhere in the world. At least 31 persons died quickly from acute radiation exposure or other injuries, and thousands of additional cancer deaths among the tens of millions of people exposed to radiation from the accident may occur during the next several decades.

According to a 2006 report by the Chernobyl Forum organized by the International Atomic Energy Agency, the primary observable health consequence of the accident was a dramatic increase in childhood thyroid cancer. The Chernobyl Forum estimated that about 4,000 cases of thyroid cancer have occurred in children who after the accident drank milk contaminated with high levels of radioactive iodine, which concentrates in the thyroid. Although the Chernobyl Forum found only 15 deaths from those thyroid cancers, it estimated that about 4,000 other cancer deaths may have occurred among the 600,000 people with the highest radiation exposures, plus an estimated 1% increase in cancer deaths among persons with less exposure. The report estimated that about 77,000 square miles were significantly contaminated by radioactive cesium.[53] Greenpeace issued a report in 2006 estimating that 200,000 deaths in Belarus, Russia, and Ukraine resulted from the Chernobyl accident between 1990 and 2004.[54]

Licensing and Regulation

For many years, a top priority of the U.S. nuclear industry was to modify the process for licensing new nuclear plants. No electric utility would consider ordering a nuclear power plant, according to the industry, unless licensing became quicker and more predictable, and designs were less subject to mid-construction safety-related changes required by NRC. The Energy Policy Act of 1992 (P.L. 102-486) largely implemented the industry's licensing goals.

Nuclear plant licensing under the Atomic Energy Act of 1954 (P.L. 83-703; U.S.C. 2011-2282) had historically been a two-stage process. NRC first issued a construction permit to build a plant and then, after construction was finished, an operating license to run it. Each stage of the licensing process involved adjudicatory proceedings. Environmental impact statements also are required under the National Environmental Policy Act.

---

[51] U.S. NRC, Regulatory Guide 1.174, "An Approach for Using Probabilistic Risk Assessment in Risk-Informed Decisions on Plant-Specific Changes to the Licensing Basis," Revision 1, November 2002, http://www.nrc.gov/ reading-rm/doc-collections/reg-guides/power-reactors/rg/01-174.

[52] Public Citizen Energy Program, "The Myth of Nuclear Safety," http://www.citizen.org/cmep/energy_enviro_nuclear/ nuclear_power_plants/reactor_safety/articles.cfm?ID=4454.

[53] The Chernobyl Forum: 2003-2005, *Chernobyl's Legacy Health, Environmental and Socio-Economic Impacts*, International Atomic Energy Agency, April 2006.

[54] Greenpeace. *The Chernobyl Catastrophe Consequences on Human Health*, April 2006, p. 10.

Over the vehement objections of nuclear opponents, the Energy Policy Act of 1992 provided a clear statutory basis for one-step nuclear licenses. Under the new process, NRC can issue combined construction permits and operating licenses (COLs) and allow completed plants to operate without delay if they meet all construction requirements—called "inspections, tests, analyses, and acceptance criteria," or ITAAC. NRC would hold preoperational hearings on the adequacy of plant construction only in specified circumstances.

DOE's Nuclear Power 2010 program had paid up to half the cost of several COLs and early site permits to test the revised licensing procedures. However, the COL process cannot be fully tested until construction of new reactors is completed. At that point, it could be seen whether completed plants will be able to operate without delays or whether adjudicable disputes over construction adequacy may arise. Section 638 of the Energy Policy Act of 2005 (EPACT05, P.L. 109-58) authorizes federal payments to the owner of a completed reactor whose operation is held up by regulatory delays. The nuclear industry is asking Congress to require NRC to use informal procedures in determining whether ITAAC have been met, eliminate mandatory hearings on uncontested issues before granting a COL, and make other changes in the licensing process.[55]

A fundamental concern in the nuclear regulatory debate is the performance of NRC in issuing and enforcing nuclear safety regulations. The nuclear industry and its supporters have regularly complained that unnecessarily stringent and inflexibly enforced nuclear safety regulations have burdened nuclear utilities and their customers with excessive costs. But many environmentalists, nuclear opponents, and other groups charge NRC with being too close to the nuclear industry, a situation that they say has resulted in lax oversight of nuclear power plants and routine exemptions from safety requirements.

Primary responsibility for nuclear safety compliance lies with nuclear plant owners, who are required to find any problems with their plants and report them to NRC. Compliance is also monitored directly by NRC, which maintains at least two resident inspectors at each nuclear power plant. The resident inspectors routinely examine plant systems, observe the performance of reactor personnel, and prepare regular inspection reports. For serious safety violations, NRC often dispatches special inspection teams to plant sites.

NRC's reactor safety program is based on "risk-informed regulation," in which safety enforcement is guided by the relative risks identified by detailed individual plant studies. NRC's risk-informed reactor oversight system, inaugurated April 2, 2000, relies on a series of performance indicators to determine the level of scrutiny that each reactor should receive.[56]

Reactor Security

Nuclear power plants have long been recognized as potential targets of terrorist attacks, and critics have long questioned the adequacy of requirements for nuclear plant operators to defend against such attacks. All commercial nuclear power plants licensed by NRC have a series of

---

[55] Nuclear Energy Institute, *Legislative Proposal to Help Meet Climate Change Goals by Expanding U.S. Nuclear Energy Production*, Washington, DC, October 28, 2009, p. 5, http://www.nei.org/resourcesandstats/documentlibrary/newplants/policybrief/2009-nuclear-policy-initiative.

[56] For more information about the NRC reactor oversight process, see http://www.nrc.gov/NRR/OVERSIGHT/ASSESS/index.html.

physical barriers against access to vital reactor areas and are required to maintain a trained security force to protect them.

A key element in protecting nuclear plants is the requirement that simulated terrorist attacks, monitored by NRC, be carried out to test the ability of the plant operator to defend against them. The severity of attacks that plant security must prepare for is specified in the "design basis threat" (DBT).

EPACT05 required NRC to revise the DBT based on an assessment of terrorist threats, the potential for multiple coordinated attacks, possible suicide attacks, and other criteria. NRC approved the DBT revision based on those requirements on January 29, 2007. The revised DBT does not require nuclear power plants to defend against deliberate aircraft attacks. NRC contended that nuclear facilities were already required to mitigate the effects of large fires and explosions, no matter what the cause, and that active protection against airborne threats was being addressed by U.S. military and other agencies.[57] After much consideration, NRC voted February 17, 2009, to require all new nuclear power plants to incorporate design features that would ensure that, in the event of a crash by a large commercial aircraft, the reactor core would remain cooled or the reactor containment would remain intact, and radioactive releases would not occur from spent fuel storage pools.[58] The rule change was published in the Federal Register June 12, 2009.[59]

NRC rejected proposals that existing reactors also be required to protect against aircraft crashes, such as by adding large external steel barriers. However, NRC did impose some additional requirements related to aircraft crashes on all reactors, both new and existing, after the 9/11 terrorist attacks of 2001. In 2002, as noted above, NRC ordered all nuclear power plants to develop strategies to mitigate the effects of large fires and explosions that could result from aircraft crashes or other causes. An NRC regulation on fire mitigation strategies, along with requirements that reactors establish procedures for responding to specific aircraft threats, was approved December 17, 2008.[60] The fire mitigation rules were published in the Federal Register March 27, 2009.[61]

Other ongoing nuclear plant security issues include the vulnerability of spent fuel pools, which hold highly radioactive nuclear fuel after its removal from the reactor, standards for nuclear plant security personnel, and nuclear plant emergency planning. NRC's March 2009 security regulations addressed some of those concerns and included a number of other security enhancements.

EPACT05 required NRC to conduct force-on-force security exercises at nuclear power plants every three years (which was NRC's previous policy), authorized firearms use by nuclear security

---

[57] NRC Office of Public Affairs, *NRC Approves Final Rule Amending Security Requirements*, News Release No. 07-012, January 29, 2007.

[58] Nuclear Regulatory Commission, *Final Rule—Consideration of Aircraft Impacts for New Nuclear Power Reactors, Commission Voting Record*, SECY-08-0152, February 17, 2009.

[59] Nuclear Regulatory Commission, "Consideration of Aircraft Impacts for New Nuclear Power Reactors," Final Rule, 74 *Federal Register* 28111, June 12, 2009. This provision is codified at 10 CFR 50.150.

[60] Nuclear Regulatory Commission, "NRC Approves Final Rule Expanding Security Requirements for Nuclear Power Plants," press release, December 17, 2008, http://www.nrc.gov/reading-rm/doc-collections/news/2008/08-227.html.

[61] Nuclear Regulatory Commission, "Power Reactor Security Requirements," Final Rule, 74 *Federal Register* 13925, March 27, 2009.

personnel (preempting some state restrictions), established federal security coordinators, and required fingerprinting of nuclear facility workers.

(For background on security issues, see CRS Report RL34331, *Nuclear Power Plant Security and Vulnerabilities*, by Mark Holt and Anthony Andrews.)

Decommissioning

When nuclear power plants reach the end of their useful lives, they must be safely removed from service, a process called *decommissioning*. NRC requires nuclear utilities to make regular contributions to dedicated funds to ensure that money is available to remove radioactive material and contamination from reactor sites after they are closed.

The first full-sized U.S. commercial reactors to be decommissioned were the Trojan plant in Oregon, whose decommissioning completion received NRC approval on May 23, 2005, and the Maine Yankee plant, for which NRC approved most of the site cleanup on October 3, 2005. The Trojan decommissioning cost $429 million, according to reactor owner Portland General Electric, and the Maine Yankee decommissioning cost about $500 million.[62] Decommissioning of the Connecticut Yankee plant cost $790 million and was approved by NRC on November 26, 2007.[63] NRC approved the cleanup of the decommissioned Rancho Seco reactor site in California on October 7, 2009.[64] The decommissioning of Rancho Seco was estimated to cost $500 million, excluding future demolition of the cooling towers and other remaining plant structures.[65]

After nuclear reactors are decommissioned, the spent nuclear fuel (SNF) accumulated during their operating lives remains stored in pools or dry casks at the plant sites. About 2,800 metric tons of spent fuel is currently stored at nine closed nuclear power plants. "Until this SNF is removed from these nine sites, the sites cannot be fully decommissioned and made available for other purposes," DOE noted in a 2008 report.[66] President Obama's decision to terminate development of an underground spent fuel repository at Yucca Mountain, NV, has increased concerns about the ultimate disposition of spent fuel at decommissioned sites. (For more information, see CRS Report R42513, *U.S. Spent Nuclear Fuel Storage*, by James D. Werner.)

Nuclear Accident Liability

Liability for damages to the general public from nuclear incidents is addressed by the Price-Anderson Act (primarily Section 170 of the Atomic Energy Act of 1954, 42 U.S.C. 2210). EPACT05 extended the availability of Price-Anderson coverage for new reactors and new DOE nuclear contracts through the end of 2025. (Existing reactors and contracts were already covered.)

---

[62] Sharp, David, "NRC Signs Off on Maine Yankee's Decommissioning," *Associated Press*, October 3, 2005.

[63] E-mail communication from Bob Capstick, Connecticut Yankee Atomic Power Company, August 28, 2008.

[64] Nuclear Regulatory Commission, "NRC Releases Rancho Seco Nuclear Plant for Unconditional Use," press release, October 7, 2009, http://www.nrc.gov/reading-rm/doc-collections/news/2009/09-165.html.

[65] "20 Years Later, Rancho Seco Ready for Final Shutdown," *Sacramento County Herald*, June 9, 2009, http://m.news10.net/news.jsp?key=190656.

[66] DOE Office of Civilian Radioactive Waste Management, Report to Congress on the Demonstration of the Interim Storage of Spent Nuclear Fuel from Decommissioned Nuclear Power Reactor Sites, DOE/RW-0596, Washington, DC, December 2008, p. 1, http://www.energy.gov/media/ES_Interim_Storage_Report_120108.pdf.

Under Price-Anderson, the owners of commercial reactors must assume all liability for nuclear damages awarded to the public by the court system, and they must waive most of their legal defenses following a severe radioactive release ("extraordinary nuclear occurrence"). To pay any such damages, each licensed reactor with at least 100 megawatts of electric generating capacity must carry the maximum liability insurance reasonably available, which was raised from $300 million to $375 million on January 1, 2010.[67] Any damages exceeding $375 million are to be assessed equally against all 100-megawatt-and-above power reactors, up to $111.9 million per reactor. Those assessments—called "retrospective premiums"—would be paid at an annual rate of no more than $17.5 million per reactor, to limit the potential financial burden on reactor owners following a major accident. According to NRC, all 104 commercial reactors are currently covered by the Price-Anderson retrospective premium requirement.[68]

For each nuclear incident, the Price-Anderson liability system currently would provide up to $12.6 billion in public compensation. That total includes the $375 million in insurance coverage carried by the reactor that suffered the incident, plus the $111.9 million in retrospective premiums from each of the 104 currently covered reactors, totaling $12.0 billion. On top of those payments, a 5% surcharge may also be imposed, raising the total per-reactor retrospective premium to $117.5 million and the total available compensation to about $12.6 billion. Under Price-Anderson, the nuclear industry's liability for an incident is capped at that amount, which varies over time depending on the number of covered reactors, the amount of available insurance, and an inflation adjustment. Payment of any damages above that liability limit would require congressional approval under special procedures in the act.

EPACT05 increased the limit on per-reactor annual payments to $15 million from the previous $10 million, and required the annual limit to be adjusted for inflation every five years. As under previous law, the total retrospective premium limit is adjusted every five years as well. Both the annual and total limits were most recently adjusted October 29, 2008.[69] For the purposes of those payment limits, a nuclear plant consisting of multiple small reactors (100-300 megawatts, up to a total of 1,300 megawatts) would be considered a single reactor. Therefore, a power plant with six 120-megawatt small modular reactors would be liable for retrospective premiums of up to $111.9 million, rather than $671.4 million (excluding the 5% surcharge).

The Price-Anderson Act also covers contractors who operate hazardous DOE nuclear facilities. EPACT05 set the liability limit on DOE contractors at $10 billion per accident, to be adjusted for inflation every five years. The first adjustment under EPACT, raising the liability limit to $11.961 billion, took effect October 14, 2009.[70] The liability limit for DOE contractors previously had been the same as for commercial reactors, excluding the 5% surcharge, except when the limit for commercial reactors dropped because of a decline in the number of covered reactors. Price-Anderson authorizes DOE to indemnify its contractors for the entire amount of their liability, so that damage payments for nuclear incidents at DOE facilities would ultimately come from the

---

[67] American Nuclear Insurers, "Need for Nuclear Liability Insurance," January 2010, http://www.nuclearinsurance.com/library/Nuclear%20Liability%20in%20the%20US.pdf.

[68] Reactors smaller than 100 megawatts must purchase an amount of liability coverage determined by NRC but are not subject to retrospective premiums. Total liability for those reactors is limited to $560 million, with the federal government indemnifying reactor operators for the difference between that amount and their liability coverage (Atomic Energy Act Sec. 170 b. and c.).

[69] Nuclear Regulatory Commission, "Inflation Adjustment to the Price-Anderson Act Financial Protection Regulations," 73 *Federal Register* 56451, September 29, 2008.

[70] Department of Energy, "Adjusted Indemnification Amount," 74 *Federal Register* 52793, October 14, 2009.

Treasury. However, the law also allows DOE to fine its contractors for safety violations, and contractor employees and directors can face criminal penalties for "knowingly and willfully" violating nuclear safety rules.

EPACT05 limited the civil penalties against a nonprofit contractor to the amount of management fees paid under that contract. Previously, Atomic Energy Act §234A specifically exempted seven nonprofit DOE contractors and their subcontractors from civil penalties and authorized DOE to automatically remit any civil penalties imposed on nonprofit educational institutions serving as DOE contractors. EPACT05 eliminated the civil penalty exemption for future contracts by the seven listed nonprofit contractors and DOE's authority to automatically remit penalties on nonprofit educational institutions.

The Price-Anderson Act's limits on liability were crucial in establishing the commercial nuclear power industry in the 1950s. Supporters of the Price-Anderson system contend that it has worked well since that time in ensuring that nuclear accident victims would have a secure source of compensation, at little cost to the taxpayer. Extension of the act was widely considered a prerequisite for new nuclear reactor construction in the United States. Opponents contend that Price-Anderson inappropriately subsidizes the nuclear power industry by reducing its insurance costs and protecting it from some of the financial consequences of the most severe conceivable accidents. The possibility that damages to the public from the Fukushima accident could greatly exceed the Price-Anderson liability limits has prompted new calls for reexamination of the law.[71]

The United States is supporting the establishment of an international liability system that, among other purposes, would cover U.S. nuclear equipment suppliers conducting foreign business. The Convention on Supplementary Compensation for Nuclear Damage (CSC) will not enter into force until at least five countries with a specified level of installed nuclear capacity have enacted implementing legislation. Such implementing language was included in the Energy Independence and Security Act of 2007 (P.L. 110-140, section 934), signed by President Bush December 19, 2007. Supporters of the Convention hope that more countries will join now that the United States has acted. Aside from the United States, three countries have submitted the necessary instruments of ratification, but the remaining nine countries that so far have signed the convention do not have the required nuclear capacity for it to take effect. Ratification by a large nuclear energy producer such as Japan would allow the treaty to take effect, as would ratification by two significant but smaller producers such as South Korea, Canada, Russia, or Ukraine.

Under the U.S. implementing legislation, the CSC would not change the liability and payment levels already established by the Price-Anderson Act. Each party to the convention would be required to establish a nuclear damage compensation system within its borders analogous to Price-Anderson. For any damages not covered by those national compensation systems, the convention would establish a supplemental tier of damage compensation to be paid by all parties. P.L. 110-140 requires the U.S. contribution to the supplemental tier to be paid by suppliers of nuclear equipment and services, under a formula to be developed by DOE. Supporters of the convention contend that it will help U.S. exporters of nuclear technology by establishing a predictable international liability system. For example, U.S. reactor sales to the growing

---

[71] Ellen Vancko, Union of Concerned Scientists, "The Impact of Fukushima on the US Nuclear Power Industry," presentation to the Center for Strategic and International Studies Conference on Nuclear Safety and Fukushima, April 7, 2011, https://csis.org/files/attachments/110407_vancko_nuclear_safety_0.pdf.

economies of China and India would be facilitated by those countries' participation in the CSC liability regime.

Federal Incentives for New Nuclear Plants

The nuclear power industry contends that support from the federal government would be needed for "a major expansion of nuclear energy generation."[72] Significant incentives for building new nuclear power plants were included in the Energy Policy Act of 2005 (EPACT05, P.L. 109-58), signed by President Bush on August 8, 2005. These include production tax credits, loan guarantees, insurance against regulatory delays, and extension of the Price-Anderson Act nuclear liability system (discussed above in the "Nuclear Accident Liability" section of this report). Relatively low prices for natural gas—nuclear power's chief competitor—and rising estimated nuclear plant construction costs have decreased the likelihood that new reactors would be built without federal support. Any regulatory delays and increased safety requirements resulting from the Fukushima accident could also pose an obstacle to nuclear construction plans.

As a result, numerous bills have been introduced in recent years to strengthen or add to the EPACT05 incentives (see "Legislation in the 112th Congress" at the end of this report). Nuclear power critics have denounced the federal support programs and proposals as a "bailout" of the nuclear industry, contending that federal efforts should focus instead on renewable energy and energy efficiency.[73]

Nuclear Production Tax Credit

EPACT05 provides a 1.8-cents/kilowatt-hour tax credit for up to 6,000 megawatts of new nuclear capacity for the first eight years of operation, up to $125 million annually per 1,000 megawatts. The credit is not adjusted for inflation.

The Treasury Department published interim guidance for the nuclear production tax credit on May 1, 2006.[74] Under the guidance, the 6,000 megawatts of eligible capacity (enough for about four or five reactors) are to be allocated among reactors that filed license applications by the end of 2008. If more than 6,000 megawatts of nuclear capacity ultimately qualify for the production tax credit, then the credit is to be allocated proportionally among any of the qualifying reactors that begin operating before 2021.

By the end of 2008, license applications had been submitted to NRC for more than 34,000 megawatts of nuclear generating capacity,[75] so if all those reactors were built before 2021 they would receive less than 20% of the maximum tax credit. However, the reactor licensing status

---

[72] Nuclear Energy Institute, "NEI Unveils Package of Policy Initiatives Needed to Achieve Climate Change Goals," press release, October 26, 2009, http://www.nei.org/newsandevents/newsreleases/nei-unveils-package-of-policy-initiatives-needed-to-achieve-climate-change-goals/.

[73] Nuclear Information and Resource Service, "Senate Appropriators Lard President Obama's Stimulus Package with up to $50 Billion in Nuclear Reactor Pork," press release, January 30, 2009, http://www.nirs.org/press/01-30-2009/1.

[74] Department of the Treasury, Internal Revenue Service, *Internal Revenue Bulletin*, No. 2006-18, "Credit for Production From Advanced Nuclear Facilities," Notice 2006-40, May 1, 2006, p. 855.

[75] Energy Information Administration, *Status of Potential New Commercial Nuclear Reactors in the United States*, February 19, 2009.

shown in **Table 1** indicates that only four new units, totaling about 4,600 megawatts of capacity, are currently licensed for construction and likely to be completed before 2021. Two other units, totaling about 2,300 megawatts, are scheduled to receive their licenses and could possibly go into service by 2021.

The Nuclear Energy Institute (NEI) has urged Congress to remove the 6,000 megawatt capacity limit for the production tax credit, index it for inflation, and extend the deadline for plants to begin operation to the start of 2025. NEI is also proposing that a 30% investment tax credit be available for new nuclear construction as an alternative to the production credit.[76]

Standby Support

Because the nuclear industry has often blamed licensing delays for past nuclear reactor construction cost overruns, EPACT05 authorizes the Secretary of Energy to provide "standby support," or regulatory risk insurance, to help pay the cost of regulatory delays at up to six new commercial nuclear reactors. For the first two reactors that begin construction, the DOE payments could cover all the eligible delay-related costs, such as additional interest, up to $500 million each. For the next four reactors, half of the eligible costs could be paid by DOE, with a payment cap of $250 million per reactor. Delays caused by the failure of a reactor owner to comply with laws or regulations would not be covered. Project sponsors will be required to pay the "subsidy cost" of the program, consisting of the estimated present value of likely future government payments.

DOE published a final rule for the "standby support" program August 11, 2006.[77] According to a DOE description of the final rule,

> Events that would be covered by the risk insurance include delays associated with the Nuclear Regulatory Commission's reviews of inspections, tests, analyses and acceptance criteria or other licensing schedule delays as well as certain delays associated with litigation in federal, state or tribal courts. Insurance coverage is not available for normal business risks such as employment strikes and weather delays. Covered losses would include principal and interest on debt and losses resulting from the purchase of replacement power to satisfy contractual obligations.[78]

Under the program's regulations, a project sponsor may enter into a conditional agreement for standby support before NRC issues a combined operating license. The first six conditional agreements to meet all the program requirements, including the issuance of a COL and payment of the estimated subsidy costs, can be converted to standby support contracts. No conditional agreements have yet been reached, according to DOE, primarily because the subsidy cost estimates have not been approved by the Office of Management and Budget (OMB).[79]

---

[76] Nuclear Energy Institute, *Legislative Proposal to Help Meet Climate Change Goals by Expanding U.S. Nuclear Energy Production*, Washington, DC, October 28, 2009, p. 4, http://www.nei.org/resourcesandstats/documentlibrary/newplants/policybrief/2009-nuclear-policy-initiative.

[77] Department of Energy, "Standby Support for Certain Nuclear Plant Delays," *Federal Register*, August 11, 2006, p. 46306.

[78] DOE press release, August 4, 2006, http://nuclear.gov/home/08-04-06.html.

[79] Meeting with Rebecca F. Smith-Kevern, Director, DOE Office of Light Water Reactor Deployment, October 7, 2009.

The Nuclear Energy Institute has called for expanding the Standby Support program to $500 million for all six covered reactors, rather than just the first two. In addition, NEI proposed that if a reactor successfully begins operating without any delay payments, that plant's Standby Support coverage, instead of expiring unused, be allowed to "roll over" to the next plant with a conditional agreement.[80]

Loan Guarantees

Title XVII of EPACT05 authorizes federal loan guarantees for up to 80% of construction costs for advanced energy projects that reduce greenhouse gas emissions, including new nuclear power plants. Under such loan guarantee agreements, the federal government would repay all covered loans if the borrower defaulted. This would reduce the risk to lenders and allow them to provide financing at low interest rates. The Title XVII loan guarantees are widely considered crucial by the nuclear industry to obtain financing for new reactors. However, opponents contend that nuclear loan guarantees would provide an unjustifiable subsidy to a mature industry and shift investment away from environmentally preferable energy technologies.[81]

The total amount of Title XVII loan guarantees to be made available for nuclear power has been the subject of considerable congressional debate. President Obama's FY2011 budget request would have nearly tripled the current ceiling on federal loan guarantees for nuclear power plants, from $18.5 billion to $54.5 billion. A $36 billion increase would increase the number of reactors that could receive loan guarantees from about three or four to about a dozen, depending on their size. The Department of Defense and Full-Year Continuing Appropriations Act for FY2011 (P.L. 112-10) did not provide the requested increase, leaving the nuclear power loan guarantee ceiling at $18.5 billion. The Administration again requested a $36 billion nuclear loan guarantee increase for FY2012, but none of the increase was included in the FY2012 Consolidated Appropriations Act. No increase was requested for FY2013.

The Administration announced the first conditional nuclear power plant loan guarantee on February 16, 2010, totaling $8.33 billion for two proposed new reactors at Georgia's Vogtle nuclear plant site. Owners of the Vogtle project have reportedly estimated that the loan guarantee could reduce their financing costs by as much as $2 billion.[82] Other finalists for the first round of nuclear reactor loan guarantees were Calvert Cliffs 3 in Maryland, South Texas Plant 3 and 4, and Summer 2 and 3.[83] However, as noted earlier, the future of the proposed units at Calvert Cliffs and the South Texas Plant is currently uncertain, leaving only Summer 2 and 3 as clearly viable candidates.

---

[80] Nuclear Energy Institute, op. cit.

[81] Thomas B. Cochran and Christopher E. Paine, *Statement on Nuclear Developments Before the Committee on Energy and Natural Resources, United States Senate*, Natural Resources Defense Council, March 18, 2009, http://energy.senate.gov/public/index.cfm?FuseAction=Hearings.Testimony&Hearing_ID=f25ddd10-c1f5-9e2e-528e-c4321cca4c1b&Witness_ID=9f14a78d-58d0-43fb-bf5b-21426d1d888e.

[82] K. Steiner-Dicks, "Weekly Intelligence Brief 7-13 June 2012," *Nuclear Energy Insider*, June 13, 2012.

[83] Letter from Office of Management and Budget Director Peter R. Orszag to House and Senate leaders, May 21, 2010, http://www.whitehouse.gov/omb/assets/legislative_letters/Pelosi_05212010.pdf.

DOE issued final rules for the program October 4, 2007,[84] and finalized the first loan guarantee on September 4, 2009, totaling $535 million to Solyndra Inc. for a photovoltaic panel manufacturing plant, which subsequently defaulted.[85] DOE's proposed loan guarantee rules, published May 16, 2007, had been sharply criticized by the nuclear industry for limiting the guarantees to 90% of a project's debt. The industry contended that EPACT05 allows all of a project's debt to be covered, as long as debt does not exceed 80% of total construction costs. In its explanation of the proposed rules, DOE expressed concern that guaranteeing 100% of a project's debt could reduce lenders' incentive to perform adequate due diligence and therefore increase default risks. In the final rule, however, DOE agreed to guarantee up to 100% of a project's debt, but in that case the loans had to be issued by the Federal Financing Bank.

Subsidy Costs

Title XVII requires the estimated future government costs resulting from defaults on guaranteed loans to be covered up-front by appropriations or by payments from project sponsors, such as the utility planning to build a plant. These "subsidy costs" are calculated as the present value of the average possible future net costs to the government for each loan guarantee. If those calculations are accurate, the subsidy cost payments for all the guaranteed projects together should cover the future costs of the program, including default-related losses. However, the Congressional Budget Office has predicted that the up-front subsidy cost payments will prove too low by at least 1% and is scoring bills accordingly.[86] For example, appropriations bills that provide loan guarantee authorizations include an adjustment equal to 1% of the loan guarantee ceiling. (For more information on loan guarantee subsidy costs, see CRS Report R42152, *Loan Guarantees for Clean Energy Technologies: Goals, Concerns, and Policy Options*, by Phillip Brown.)

DOE loan guarantees for renewable energy and electricity transmission projects under EPACT05 section 1705, added by the American Recovery and Reinvestment Act of 2009 (P.L. 111-5), do not require subsidy cost payments by project sponsors, because potential losses are covered by advance appropriations in the act. No such appropriations are currently available for nuclear power projects, so it is anticipated that nuclear loan guarantee subsidy costs would be paid by the project sponsors. As a result, the level of the subsidy costs could have a powerful effect on the viability of nuclear power projects, which are currently expected to cost between $5 billion and $10 billion per reactor. For example, a 10% subsidy cost for a $7 billion loan guarantee would require an up-front payment of $700 million.

No subsidy cost amount has yet been established for any nuclear loan guarantee, including the lead Vogtle project in Georgia. The Administration's continuing internal deliberations over that question may reflect its importance and the amount of controversy being generated. Internal DOE documents released May 23, 2012, pursuant to the Freedom of Information Act show that Southern Company, the lead partner in the Vogtle project, has been offered a subsidy cost of

---

[84] Published October 23, 2007 (72 *Federal Register* 60116). Revised final rules were published December 4, 2009 (74 *Federal Register* 63544).

[85] Department of Energy, "Vice President Biden Announces Finalized $535 Million Loan Guarantee," press release, September 4, 2009, http://www.lgprogram.energy.gov/press/090409.pdf. For details on the default, see CRS Report R42058, *Market Dynamics That May Have Contributed to Solyndra's Bankruptcy*, by Phillip Brown.

[86] Congressional Budget Office, *S. 1321, Energy Savings Act of 2007*, CBO Cost Estimate, Washington, DC, June 11, 2007, pp. 7-9, http://www.cbo.gov/ftpdocs/82xx/doc8206/s1321.pdf.

0.5%-1.5%, subject to other conditions that are still under negotiation. Higher subsidy costs are being offered to two other partners in the project.[87]

The nuclear industry contends that historical experience indicates defaults are likely to be minimal and that nuclear plant subsidy costs should therefore be low.[88] However, nuclear power critics contend that nuclear power plants are likely to experience delays and cost overruns that could lead to much larger losses under the loan guarantee program. The Center for American Progress concluded that nuclear subsidy costs "should be at least 10 percent and possibly much more."[89]

Constellation Energy informed DOE on October 8, 2010, that it was withdrawing from loan guarantee negotiations on Calvert Cliffs 3, blaming "the Office of Management and Budget's inability to address significant problems with its methodology for determining the project's credit subsidy cost." Constellation's letter to DOE said OMB's "shockingly high" estimate of the subsidy cost for Calvert Cliffs 3 was 11.6%, or about $880 million. "Such a sum would clearly destroy the project's economics (or the economics of any nuclear project for that matter), and was dramatically out of line with both our own and independent assessments of what the figure should reasonably be," the letter stated.[90] Although OMB has not released its subsidy cost methodology, it may consider the default risk for a "merchant plant" such as Calvert Cliffs to be significantly higher than that of a rate-regulated plant such as Vogtle. A plant under traditional rate regulation is allowed to pass all prudently incurred costs through to utility ratepayers, while a merchant plant charges market rates for its power. A merchant plant, therefore, could potentially earn higher profits than a rate-regulated plant, but it also runs the risk of being unable to cover its debt payments if market rates for wholesale electric power drop too low or if its costs are higher than anticipated.

Congressionally Authorized Ceilings

Under the Federal Credit Reform Act (FCRA), federal loan guarantees cannot be provided without an authorized level in an appropriations act. The Senate-passed version of omnibus energy legislation in the 110th Congress (H.R. 6) would have explicitly eliminated FCRA's applicability to DOE's planned loan guarantees under EPACT05 (Section 124(b)). That provision would have given DOE essentially unlimited loan guarantee authority for guarantees whose subsidy costs were paid by project sponsors, but it was dropped from the final legislation (P.L. 110-140). Similar language was also included in subsequent legislative proposals, such as energy legislation reported by the Senate Committee on Energy and Natural Resources July 16, 2009 (S. 1462).

---

[87] Southern Alliance for Clean Energy, "Secret Documents Highlight Nuclear's Risk," press release, May 23, 2012, http://www.cleanenergy.org/index.php?/Press-Update.html?form_id=8&item_id=299.

[88] Statement of Leslie C. Kass, Nuclear Energy Institute, to the Subcommittee on Domestic Policy, House Committee on Oversight and Government Reform, April 20, 2010, http://www.nei.org/newsandevents/speechesandtestimony/april-20-2010-kass. DOE is treating final subsidy cost determinations as proprietary, prompting some groups to call for the amounts to be made public.

[89] Richard Caperton, *Protecting Taxpayers from a Financial Meltdown*, Center for American Progress, Washington, DC, March 8, 2010, p. 2, http://www.americanprogress.org/issues/2010/03/nuclear_financing.html.

[90] Letter from Michael J. Wallace, Vice Chairman and Chief Operating Officer, Constellation Energy, to Dan Poneman, Deputy Secretary of Energy, October 8, 2010, http://media.washingtonpost.com/wp-srv/hp/ssi/wpc/constellationenergy.PDF?sid=ST2010100900005.

Pursuant to FCRA, the FY2007 continuing resolution (P.L. 110-5) established an initial cap of $4 billion on loan guarantees under the program, without allocating that amount among the various eligible technologies. The explanatory statement for the FY2008 omnibus funding act (P.L. 110-161) increased the loan guarantee ceiling to $38.5 billion through FY2009, including $18.5 billion specifically for nuclear power plants and $2 billion for uranium enrichment plants.[91]

The FY2009 omnibus funding act increased DOE's total loan guarantee authority for specified technology categories to $47 billion, in addition to the $4 billion in general authority provided in FY2007. Of the $47 billion, $18.5 billion continued to be reserved for nuclear power, $18.5 billion was for energy efficiency and renewables, $6 billion was for coal, $2 billion was for carbon capture and sequestration, and $2 billion was for uranium enrichment. The time limits on the loan guarantee authority were eliminated. The loan guarantee ceilings remained the same for FY2010 but were sharply reduced for non-nuclear technologies by the FY2011 Continuing Appropriations Act. The nuclear power loan guarantee ceiling remains at $18.5 billion.

Nuclear Solicitations

DOE issued a solicitation for up to $20.5 billion in nuclear power and uranium enrichment plant loan guarantees on June 30, 2008.[92] According to the nuclear industry, 10 nuclear power projects applied for $93.2 billion in loan guarantees, and two uranium enrichment projects asked for $4.8 billion in guarantees, several times the amount available.[93] Under the program's regulations, a conditional loan guarantee commitment cannot become a binding loan guarantee agreement until the project receives a COL and all other regulatory requirements are met, as noted above; and the first COLs were issued in early 2012.

In the uranium enrichment solicitation, DOE in July 2009 informed USEC Inc., which plans to build a new plant in Ohio, that its technology needed further testing before a loan guarantee could be issued.[94] DOE notified Congress in March 2010 that it would reprogram $2 billion of its unused FY2007 loan guarantee authority toward uranium enrichment, increasing the uranium enrichment total to $4 billion. The move would potentially allow guarantees to be provided to both USEC and the other applicant in the uranium enrichment solicitation, the French firm Areva, which is planning a plant in Idaho.[95] DOE offered a $2 billion conditional loan guarantee to Areva on May 20, 2010.[96]

DOE informed USEC in October 2011 that the centrifuge technology for its proposed new enrichment plant still needed further testing and offered to provide up to $300 million to help build a demonstration "train" of 720 centrifuges.[97] Energy Secretary Steven Chu sent letters to the

---

[91] *Congressional Record*, December 17, 2007, p. H15585.

[92] http://www.lgprogram.energy.gov/keydocs.html.

[93] Marvin S. Fertel, *Statement for the Record to the Committee on Energy and Natural Resources, U.S. Senate*, Nuclear Energy Institute, March 18, 2009, p. 9, http://energy.senate.gov/public/index.cfm?FuseAction=Hearings.Testimony& Hearing_ID=f25ddd10-c1f5-9e2e-528e-c4321cca4c1b&Witness_ID=4de5e2df-53fe-49ba-906e-9b69d3674e41.

[94] Department of Energy, "800 to 1000 New Jobs Coming to Piketon," press release, July 28, 2009, http://www.lgprogram.energy.gov/press/072809.pdf.

[95] Maureen Conley, "DOE Finds $2 Billion More for SWU Plant Loan Guarantees," *NuclearFuel*, April 5, 2010, p. 3.

[96] Department of Energy, "DOE Offers Conditional Loan Guarantee for Front End Nuclear Facility in Idaho," press release, May 20, 2010, http://www.energy.gov/news/8996.htm.

[97] Maureen Conley, "Congress 'Frustrated' by Inaction on USEC Loan Guarantee," *NuclearFuel*, October 31, 2011, p. (continued...)

House and Senate Appropriations Committees on October 25, 2011, to request an unspecified funding transfer in FY2012 for the first $150 million of USEC assistance.[98] DOE's FY2013 budget request includes $150 million for the USEC centrifuge demonstration program. The House provided $100 million in the FY2013 Energy and Water Development Appropriations Bill (H.R. 5325, H.Rept. 112-462), while the Senate Appropriations Committee version of the bill recommended $150 million in transfer authority to fund the project (S. 2465, S.Rept. 112-164). An authorization of $150 million for the USEC centrifuge demonstration program is included in the House-passed National Defense Authorization Act for Fiscal Year 2013 (H.R. 4310).

DOE has recently provided other assistance to USEC. DOE agreed on May 15, 2012, to provide depleted uranium stockpiles (material left over from the enrichment process) to Energy Northwest for reenrichment at USEC's plant in Paducah, KY, for use as reactor fuel.[99] DOE agreed on March 13, 2012, to acquire low-enriched uranium from USEC in exchange for taking responsibility for low-value depleted uranium tails that USEC would otherwise have to dispose of, freeing $44 million of USEC's funds for the centrifuge project. [100] DOE announced June 13, 2012, that it would provide $88 million for the centrifuge demonstration program by taking over responsibility for disposal of additional depleted uranium from USEC. In return, DOE will take ownership of the equipment and technology used in the demonstration and lease it to USEC.[101]

Global Climate Change

Global climate change that may be caused by carbon dioxide and other greenhouse gas emissions is cited by nuclear power supporters as an important reason to develop a new generation of reactors. Nuclear power plants emit relatively little carbon dioxide, mostly from nuclear fuel production and auxiliary plant equipment. This "green" nuclear power argument has received growing attention in think tanks and academia. As stated by the Massachusetts Institute of Technology in its major study *The Future of Nuclear Power*: "Our position is that the prospect of global climate change from greenhouse gas emissions and the adverse consequences that flow from these emissions is the principal justification for government support of the nuclear energy option."[102] As discussed above, the Obama Administration is also including nuclear power as part of its clean energy strategy.

(...continued)

8.

[98] Steven Chu, Secretary of Energy, letters to Chairmen and Ranking Members of House and Senate Appropriations Committees and Subcommittees on Energy and Water Development, October 25, 2011. For more information on the USEC funding proposal, see CRS Congressional Distribution Memorandum *Business Outlook for USEC Inc.*, by Mark Holt, available from the author.

[99] USEC Inc., "Five-Party Arrangement Extends Paducah Gaseous Diffusion Plant Enrichment Operations," press release, May 15, 2012, http://www.usec.com/news/five-party-arrangement-extends-paducah-gaseous-diffusion-plant-enrichment-operations. The depleted uranium consists of "high assay" tails, which have relatively high levels of fissile U-235.

[100] USEC Inc., "Funding," web page, http://www.usec.com/american-centrifuge/what-american-centrifuge/plant/funding.

[101] Department of Energy, "Obama Administration Announces Major Step Forward for the American Centrifuge Plant," press release, June 13, 2012, http://energy.gov/articles/obama-administration-announces-major-step-forward-american-centrifuge-plant.

[102] Interdisciplinary MIT Study, *The Future of Nuclear Power*, Massachusetts Institute of Technology, 2003, p. 79.

However, environmental groups have contended that nuclear power's potential greenhouse gas benefits are modest and must be weighed against the technology's safety risks, its potential for nuclear weapons proliferation, and the hazards of radioactive waste.[103] They also contend that energy efficiency and renewable energy would be far more productive investments for reducing greenhouse gas emissions.[104]

Proposals to reduce carbon dioxide emissions – through taxation, a cap-and-trade system, or other regulatory controls – could significantly increase the cost of generating electricity with fossil fuels and improve the competitive position of nuclear power. A federal Clean Energy Standard that includes nuclear power, as proposed in President Obama's January 2011 State of the Union Address and in S. 2146, could provide a similar boost to nuclear energy expansion. Utilities that have applied for nuclear power plant licenses have often cited the possibility of federal greenhouse gas controls or other mandates as one of the reasons for pursuing new reactors. (For more on federal incentives and the economics of nuclear power and other electricity generation technologies, see CRS Report RL34746, *Power Plants: Characteristics and Costs*, by Stan Mark Kaplan.)

Nuclear Power Research and Development

The Obama Administration's FY2013 funding request for nuclear energy research and development totaled $770.4 million. Including advanced reactors, fuel cycle technology, infrastructure support, and safeguards and security, the total nuclear energy request was $88.3 million (10%) below the enacted FY2012 funding level. Funding for safeguards and security at DOE's Idaho facilities in FY2012 was provided under a separate appropriations account, Other Defense Activities, but it was included under the Nuclear Energy account in the FY2013 request. The largest proposed reductions for FY2013 were Reactor Concepts (36%), Radiological Facility Management (27%), and Nuclear Energy Enabling Technologies (13%).

Nuclear energy funding is included in the Energy and Water Development appropriations bills. The House passed its version of the Energy and Water bill for FY2013 on June 6, 2012 (H.R. 5325, H.Rept. 112-462). Excluding funding for Idaho safeguards and security, the House bill provided an increase of $89.9 million for the nuclear energy account, for a total of $765.4 million. The House bill included $93.4 million for Idaho safeguards and security under the Other Defense Activities Account. The Senate Appropriations Committee on April 26, 2012, recommended a $20.1 million increase for nuclear energy, including Idaho safeguards and security and $17.7 million in prior-year balances (S. 2465, S.Rept. 112-164).

Using reorganized budget categories established for FY2011, the Administration's FY2013 nuclear R&D budget request is consistent with DOE's *Nuclear Energy Research and*

---

[103] Gronlund, Lisbeth, David Lochbaum, and Edwin Lyman, *Nuclear Power in a Warming World*, Union of Concerned Scientists, December 2007.

[104] Travis Madsen, Tony Dutzik, and Bernadette Del Chiaro, et al., *Generating Failure How Building Nuclear Power Plants Would Set America Back in the Race Against Global Warming*, Environment America Research and Policy Center, November 2009, http://www.environmentamerica.org/uploads/39/62/3962c378b66c4552624d09cbd8ebba02/Generating-Failure—Environment-America—Web.pdf.

Development Roadmap issued in April 2010.[105] The Roadmap lays out the following four main goals for the program:

- Develop technologies and other solutions that can improve the reliability, sustain the safety, and extend the life of current reactors;

- Develop improvements in the affordability of new reactors to enable nuclear energy to help meet the Administration's energy security and climate change goals;

- Develop sustainable nuclear fuel cycles; and

- Understand and minimize the risks of nuclear proliferation and terrorism.

Reactor Concepts

The Reactor Concepts program area includes the Next Generation Nuclear Plant (NGNP) demonstration project and research on other advanced reactors (often referred to as Generation IV reactors). This area also includes funding for developing advanced small modular reactors (discussed in the next section) and to enhance the "sustainability" of existing commercial light water reactors. The total FY2013 funding request for this program was $73.7 million, a reduction of $41.2 million from FY2012. The House provided an increase of $11.1 million from the FY2012 level, while the Senate Appropriations Committee's recommendation was the same as the request.

Most of the Administration's proposed reduction in Reactor Concepts would be for NGNP, a high-temperature gas-cooled reactor demonstration project authorized by the Energy Policy Act of 2005 (EPACT05, P.L. 109-58). The reactor is intended to produce high-temperature heat that could be used to generate electricity, help separate hydrogen from water, or be used in other industrial processes. DOE is requesting $21.2 million for the NGNP project for FY2013, down from $40 million provided in FY2012. Under EPACT05, the Secretary of Energy was to decide by the end of FY2011 whether to proceed toward construction of a demonstration plant. Secretary of Energy Steven Chu informed Congress on October 17, 2011, that DOE would not proceed with a demonstration plant design "at this time" but would continue research on the technology.[106] Potential obstacles facing NGNP include low prices for natural gas, the major competing fuel, and private-sector unwillingness to share the project's costs as required by EPACT05.[107] According to the DOE budget justification, the NGNP program in FY2013 will focus on fuels for very high temperature reactors, the graphite used in high-temperature reactor cores, and licensing issues. The House provided $50 million for NGNP, to allow DOE to continue developing a licensing framework and continue working with industry on the program. The Senate panel restricted NGNP activities to ongoing fuel-related research.

Funding for the Advanced Reactor Concepts subprogram would also be reduced sharply by the Administration request, from $21.9 million in FY2012 to $12.4 million in FY2013. Reactor

---

[105] Department of Energy, *Nuclear Energy Research and Development Roadmap*, Report to Congress, Washington, DC, April 2010, http://nuclear.gov/pdfFiles/NuclearEnergy_Roadmap_Final.pdf.

[106] Idaho National Laboratory, *NGNP Project 2011 Status and Path Forward*, INL/EXT-11-23907, December 2011.

[107] Yanmei Xie, "Cheap Natural Gas, Cost-Share Disagreement Jeopardize NGNP," *Nucleonics Week*, April 28, 2011, p. 1.

---

concepts being developed by this subprogram are generally classified as "Generation IV" reactors, as opposed to the existing fleet of commercial light water reactors, which are generally classified as generations II and III. Such advanced reactors "could dramatically improve nuclear power performance including sustainability, economics, and safety and proliferation resistance," according to the FY2013 justification. Nuclear technology development under this program includes "fast reactors," using high-energy neutrons, and reactors that would use a variety of heat-transfer fluids, such as liquid sodium and supercritical carbon dioxide. International research collaboration in this area would continue under the Generation IV International Forum (GIF). The House provided an increase of $1.1 million over FY2012, while the Senate Appropriations Committee approved the Administration's proposed reduction.

DOE's FY2013 request for the Light Water Reactor Sustainability subprogram was $21.7 million, $3.3 million below the FY2012 appropriation. The program conducts research on extending the life of existing commercial light water reactors beyond 60 years, the maximum operating period currently licensed by the Nuclear Regulatory Commission. The program, which is to be cost-shared with the nuclear industry, is to study the aging of reactor materials and analyze safety margins of aging plants. Other research under this program is to focus on improving the efficiency of existing plants, through such measures as increasing plant capacity and upgrading instrumentation and control systems. Research on longer-life LWR fuel is aimed at eliminating radioactive leakage from nuclear fuel and increasing its accident tolerance, along with other "post-Fukushima lessons learned research needs," according to the budget justification. The House rejected the Administration's proposed reduction, while the Senate Appropriations Committee approved it.

Small Modular Light Water Reactors

Rising cost estimates for large conventional nuclear reactors—widely projected to be $6 billion or more—have contributed to growing interest in proposals for small modular reactors (SMRs). Ranging from about 40 to 350 megawatts of electrical capacity, such reactors would be only a fraction of the size of current commercial reactors. Several modular reactors would be installed together to make up a power block with a single control room, under most concepts. Current SMR proposals would use a variety of technologies, including high-temperature gas technology in the NGNP program and the light water (LWR) technology used by today's commercial reactors.

DOE requested $65 million for FY2013 to provide technical support for licensing small modular LWRs, $2 million below the FY2012 funding level. This program focuses on LWR designs because they are believed most likely to be deployed in the near term, according to DOE. Conferees on the FY2012 appropriations bill anticipated a five-year program totaling $452 million. The program is similar to DOE's support for larger commercial reactor designs under the Nuclear Power 2010 Program, which ended in FY2010. DOE will provide support for design certification, standards, and licensing. As with the Nuclear Power 2010 Program, at least half the costs of the LWR SMR program are to be covered by industry partners, according to DOE. The program will support two teams of reactor vendors and specific utilities or consortia who are interested in building the reactors at specific sites, according to the DOE justification. DOE announced a funding solicitation for the program on March 22, 2012.[108] Applications have been

---

[108] Department of Energy, "Obama Administration Announces $450 Million to Design and Commercialize U.S. Small Modular Nuclear Reactors," press release, March 22, 2012, http://www.ne.doe.gov/newsroom/2012PRs/ (continued...)

submitted by four industry consortia, led by Babcock & Wilcox, Holtec, NuScale Power, and Westinghouse, proposing reactors ranging from 45-225 megawatts.[109]

The House approved $114 million for the SMR licensing program, $47 million above FY2012. The House Appropriations Committee report called the increase necessary to keep the program on track to receive $452 million over five years. The Senate panel provided the same funding as in the budget request.

An additional $18.5 million for FY2013 was requested by DOE under the Reactor Concepts program (described in the section above) for SMR advanced concepts R&D—$10.2 million below the FY2012 funding level. Unlike the SMR licensing support program, which focuses on conventional LWR technology, the SMR advanced concepts program would conduct research on technologies that might be deployed in the longer term, according to the budget justification. The House rejected the Administration's proposed reduction, while the Senate Appropriations Committee approved the budget request.

Small modular reactors would go against the overall trend in nuclear power technology toward ever-larger reactors intended to spread construction costs over a greater output of electricity. Proponents of small reactors contend that they would be economically viable despite their far lower electrical output because modules could be assembled in factories and shipped to plant sites, and because their smaller size would allow for simpler safety systems. In addition, although modular plants might have similar or higher costs per kilowatt-hour than conventional large reactors, their ability to be constructed in smaller increments could reduce electric utilities' financial commitment and risk.

Fuel Cycle Research and Development

The Fuel Cycle Research and Development Program conducts "long-term, science-based" research on a wide variety of technologies for improving the management of spent nuclear fuel, according to the DOE budget justification. The total FY2013 funding request for this program is $175.4 million, $10.8 million below the FY2012 appropriation. The House approved $138.7 million for Fuel Cycle R&D, $36.7 million below the request. The Senate Appropriations Committee recommended $193.1 million, $17.7 million above the request.

The range of fuel cycle technologies being studied by the program includes direct disposal of spent fuel (the "once through" cycle) and partial and full recycling, according to the budget justification. The Fuel Cycle R&D Program "will research and develop a suite of technology options that will enable future decision-makers to make informed decisions about how best to manage nuclear waste and used fuel from reactors," the budget justification says.

Much of the planned research on spent fuel management options will address the near-term recommendations of the Blue Ribbon Commission on America's Nuclear Future, which issued its final report on January 26, 2012.[110] The commission was chartered to develop alternatives to the

---

(...continued)

nePR032212_print.html.

[109] World Nuclear Association, "Small Nuclear Power Reactors," May 2012, http://www.world-nuclear.org/info/inf33.html.

[110] Blue Ribbon Commission on America's Nuclear Future, "Blue Ribbon Commission on America's Nuclear Future (continued...)

planned Yucca Mountain, NV, spent fuel repository, which President Obama wants to terminate. The largest subprogram under Fuel Cycle Research and Development is Used Nuclear Fuel Disposition, with a request of $59.7 million, the same as the FY2012 funding level. Activities in that area include work toward the development and licensing of standardized spent fuel containers, studies of potential spent fuel disposal partnerships, and the accelerated characterization of potential geologic media for waste disposal.

The House report contended that much of the proposed research in the Used Fuel Disposition Program relates to waste program changes recommended by the Blue Ribbon Commission that have not been enacted by Congress. As a result, the panel reduced funding for Used Fuel Disposition to $38 million, $15 million of which would be for storage and transportation work related to the Yucca Mountain repository. The Senate panel's $17.7 million increase from the budget request consists of prior-year funds that would be used for a spent fuel storage pilot project (see the "Nuclear Waste Management" section for more details).

Other major research areas in the Fuel Cycle R&D Program include the development of advanced fuels for existing commercial reactors and advanced reactors, improvements in nuclear waste characteristics, and technology to increase nuclear fuel resources, such as uranium extraction from seawater.

Nuclear Energy Enabling Technologies

The Nuclear Energy Enabling Technologies (NEET) program "is designed to conduct research and development (R&D) in crosscutting technologies that directly support and enable the development of new and advanced reactor designs and fuel cycle technologies," according to the FY2013 DOE budget justification. The DOE funding request for the program was $65.3 million, $9.4 million below the FY2012 level. The House provided $75 million, nearly the same as in FY2012, while the Senate Appropriations Committee recommended the same funding as the request.

DOE's proposed funding cut would come entirely under the category of Crosscutting Technology Development, for which $26.2 million was requested, $9.7 million below FY2012. According to the budget justification, the cuts result from elimination of research on manufacturing methods and nonproliferation risk assessments. Continuing crosscutting research activities are to include development of innovative materials, advanced automation and information technologies, advanced sensors, and improved fuel performance. The Energy Innovation Hub for Modeling and Simulation, created in FY2010, had a request of $24.6 million, slightly above the FY2012 appropriation. The Modeling and Simulation Hub is creating a computer model of an operating reactor to allow a better understanding of nuclear technology, with the benefits of such modeling extending to other energy technologies in the future, according to the budget justification.

DOE requested $14.6 million for the National Scientific User Facility, the same as the FY2012 appropriation, to support partnerships by universities and other research organizations to conduct experiments "at facilities not normally accessible to these organizations," according to the

---

(...continued)

Issues Final Report to Secretary of Energy," press release, January 26, 2012, http://brc.gov/index.php?q= announcement/brc-releases-their-final-report.

justification. Up to five such partnerships are currently anticipated, and the FY2013 funding will allow up to three new long-term and five "rapid turnaround" projects to be awarded.

Nuclear Waste Management

One of the most controversial aspects of nuclear power is the disposal of radioactive waste, which can remain hazardous for thousands of years. Each nuclear reactor produces an annual average of about 20 metric tons of highly radioactive spent nuclear fuel, for a nationwide total of about 2,000 metric tons per year. U.S. reactors also generate about 27,000 cubic meters of low-level radioactive waste per year, including contaminated components and materials resulting from reactor decommissioning.[111]

The federal government is responsible for permanent disposal of commercial spent fuel (paid for with a fee on nuclear power production) and federally generated radioactive waste, while states have the authority to develop disposal facilities for most commercial low-level waste. Under the Nuclear Waste Policy Act (NWPA, 42 U.S.C. 10101, et seq.), spent fuel and other highly radioactive waste is to be isolated in a deep underground repository, consisting of a large network of tunnels carved from rock that has remained geologically undisturbed for hundreds of thousands of years. As amended in 1987, NWPA designated Yucca Mountain in Nevada as the only candidate site for the national repository. The act required DOE to begin taking waste from nuclear plant sites by 1998—a deadline that even under the most optimistic scenarios will be missed by more than 20 years. DOE filed a license application with NRC for the proposed Yucca Mountain repository in June 2008.

The Obama Administration "has determined that developing the Yucca Mountain repository is not a workable option and the Nation needs a different solution for nuclear waste disposal," according to the DOE FY2011 budget justification. As a result, no funding for Yucca Mountain or DOE's Office of Civilian Radioactive Waste Management (OCRWM), which had run the program, was requested for FY2011. The Continuing Appropriations Act for FY2011 (P.L. 112-10) approved the funding termination. The Administration established the Blue Ribbon Commission on America's Nuclear Future on March 1, 2010, to develop an alternative waste management strategy.

DOE filed a motion with NRC to withdraw the Yucca Mountain license application on March 3, 2010. An NRC licensing panel rejected DOE's withdrawal motion June 29, 2010, on the grounds that NWPA requires full consideration of the license application by NRC. The full NRC Commission deadlocked on the issue September 9, 2011, leaving the licensing panel's decision in place and prohibiting DOE from withdrawing the Yucca Mountain application. However, the commission ordered at the same time that the licensing process be halted because of "budgetary limitations."[112] No funding was provided in FY2012 or requested for FY2013 to continue Yucca

---

[111] DOE, Manifest Information Management System http://mims.apps.em.doe.gov. Average annual utility disposal from 2002 through 2011. Annual volume ranges from 68,441 cubic meters in 2005 to 5,326 cubic meters in 2009.

[112] Nuclear Regulatory Commission, "In the Matter of U.S. Department of Energy (High-Level Waste Repository)," CLI-11-07, September 9, 2011, http://www.nrc.gov/reading-rm/doc-collections/commission/orders/2011/2011-07cli.pdf.

Mountain licensing activities, although the issue is currently the subject of a federal appeals court case.[113]

The Blue Ribbon Commission issued its final report on January 26, 2012.[114] The commission recommended options for temporary storage, treatment, and permanent disposal of highly radioactive nuclear waste, along with an evaluation of nuclear waste research and development programs and the need for legislation. It did not recommend specific sites for new nuclear waste facilities or evaluate the suitability of Yucca Mountain.

The commission's proposed "consent-based" approach to the siting of waste facilities called for the roles of local, state, and tribal governments to be negotiated for each potential site. The development of consolidated waste storage and disposal facilities should begin as soon as possible, the commission urged. A new waste management organization should be established to develop the repository, along with associated transportation and storage systems, according to the commission. The new organization should have "assured access" to the Nuclear Waste Fund, which holds fees collected from nuclear power plant operators to pay for waste disposal. Under NWPA, DOE could not spend those funds without congressional appropriations.

In the FY2013 Energy and Water Development appropriations bill (H.R. 5325), the House Appropriations Committee sharply criticized the Administration's nuclear waste policy and provided $25 million for DOE to resume work on the Yucca Mountain repository license. An amendment on the House floor provided an additional $10 million to NRC for Yucca Mountain licensing (H.Amdt. 1188). The Senate Appropriations Committee provided no funds for Yucca Mountain but included language (§312, S. 2465) authorizing a pilot program to demonstrate one or more consolidated interim storage facilities for spent nuclear fuel and high level waste. Any proposed storage site would require the consent of the affected state governor, local government of jurisdiction, affected Indian tribes, and Congress. The Senate panel directed DOE to use $2 million of its program direction funding for the pilot program, along with $17.7 million in unobligated prior-year appropriations from the Nuclear Waste Fund.

Funding for the nuclear waste program in the past has been provided under two appropriations accounts. The Administration's last request for funding, in FY2010, was divided evenly between an appropriation from the Nuclear Waste Fund, which holds fees paid by nuclear utilities, and the Defense Nuclear Waste Disposal account, which pays for disposal of high-level waste from the nuclear weapons program. The Senate Appropriations Committee report for that year called for the Secretary of Energy to suspend fee collections, "given the Administration's decision to terminate the Yucca Mountain repository program while developing disposal alternatives," but the language was dropped in conference. Energy Secretary Steven Chu in October 2009 rejected requests from the nuclear industry and state utility regulators to suspend the fee, saying the revenues were still necessary, and nuclear utilities and regulators filed lawsuits to stop the fee in April 2010.[115] The U.S. Court of Appeals for the District of Columbia Circuit agreed with the

---

[113] U.S. Circuit Court of Appeals for the District of Columbia Circuit, USCA Case #11-1271, Yucca Mountain Reply Brief of Petitioners Mandamus Action, February 13, 2012, http://www.naruc.org/policy.cfm?c=filings.

[114] Blue Ribbon Commission on America's Nuclear Future, *Report to the Secretary of Energy*, January 2012, http://brc.gov/sites/default/files/documents/brc_finalreport_jan2012.pdf.

[115] National Association of Regulatory Utility Commissioners, "State Regulators Go to Court with DOE over Nuclear Waste Fees," news release, April 2, 2010, http://www.naruc.org/News/default.cfm?pr=193; Nuclear Energy Institute, "NEI, Electric Utilities File Suit to Suspend Collection of Fees for Reactor Fuel Management," news release, April 5, 2010, http://www.nei.org/newsandevents/newsreleases/nei-electric-utilities-file-suit-to-suspend-collection-of-fee-for- (continued...)

plaintiffs on June 1, 2012, and ordered DOE to prepare a new justification for continuing to collect the fees.[116]

The Yucca Mountain project faced regulatory uncertainty even before the Obama Administration's move to shut it down. A ruling on July 9, 2004, by the U.S. Court of Appeals for the District of Columbia Circuit overturned a key aspect of the Environmental Protection Agency's (EPA's) regulations for the planned repository.[117] The three-judge panel ruled that EPA's 10,000-year compliance period was too short, but it rejected several other challenges to the rules. EPA published new standards on October 15, 2008, that would allow radiation exposure from the repository to increase after 10,000 years.[118] The State of Nevada has filed a federal Appeals Court challenge to the EPA standards. (For more information on the EPA standards, see CRS Report RL34698, *EPA's Final Health and Safety Standard for Yucca Mountain*, by Bonnie C. Gitlin.)

NWPA required DOE to begin taking waste from nuclear plant sites by January 31, 1998. Nuclear utilities, upset over DOE's failure to meet that deadline, have won two federal court decisions upholding the department's obligation to meet the deadline and to compensate utilities for any resulting damages. Utilities have also won several cases in the U.S. Court of Federal Claims. DOE estimates that liability payments would eventually total $20.8 billion if DOE were to begin removing waste from reactor sites by 2020, the previous target for opening Yucca Mountain.[119] (For more information, see CRS Report R40996, *Contract Liability Arising from the Nuclear Waste Policy Act (NWPA) of 1982*, by Todd Garvey CRS Report R40202, *Nuclear Waste Disposal: Alternatives to Yucca Mountain*, by Mark Holt, CRS Report RL33461, *Civilian Nuclear Waste Disposal*, by Mark Holt, and CRS Report R42513, *U.S. Spent Nuclear Fuel Storage*, by James D. Werner.)

Nuclear Weapons Proliferation

Renewed interest in nuclear power throughout the world has led to increased concern about nuclear weapons proliferation, because technology for making nuclear fuel can also be used to produce nuclear weapons material. Of particular concern are uranium enrichment, a process to separate and concentrate the fissile isotope uranium-235, and nuclear spent fuel reprocessing, which can produce weapons-useable plutonium.

The International Atomic Energy Agency (IAEA) conducts a safeguards program that is intended to prevent civilian nuclear fuel facilities from being used for weapons purposes, but not all potential weapons proliferators belong to the system, and there are ongoing questions about its effectiveness. Several proposals have been developed to guarantee nations without fuel cycle

---

(...continued)

reactor-fuel-management.

[116] U.S. Court of Appeals for the District of Columbia Circuit, *National Association of Regulatory Utility Commissioners v. U.S. Department of Energy*, No. 11-1066, decided June 1, 2012, http://www.cadc.uscourts.gov/internet/opinions.nsf/4B11622F4FF75FEC85257A100050A681/$file/11-1066-1376508.pdf.

[117] U.S. Court of Appeals for the District of Columbia Circuit, *Nuclear Energy Institute v. Environmental Protection Agency*, No. 01-1258, July 9, 2004.

[118] Environmental Protection Agency, "Public Health and Environmental Radiation Protection Standards for Yucca Mountain, Nevada," 73 *Federal Register* 61256, October 15, 2008.

[119] BRC Final Report, op. cit., p. 80.

---

facilities a supply of nuclear fuel in exchange for commitments to forgo enrichment and reprocessing, which was one of the original goals of the Bush Administration's Global Nuclear Energy Partnership, now called the International Framework for Nuclear Energy Cooperation.[120]

Several situations have arisen throughout the world in which ostensibly commercial uranium enrichment and reprocessing technologies have been subverted for military purposes. In 2003 and 2004, it became evident that Pakistani nuclear scientist A.Q. Khan had sold sensitive technology and equipment related to uranium enrichment to states such as Libya, Iran, and North Korea. Although Pakistan's leaders maintain they did not acquiesce in or abet Khan's activities, Pakistan remains outside the Nuclear Nonproliferation Treaty (NPT) and the Nuclear Suppliers Group (NSG). Iran has been a direct recipient of Pakistani enrichment technology.

IAEA's Board of Governors found in 2005 that Iran's breach of its safeguards obligations constituted noncompliance with its safeguards agreement, and referred the case to the U.N. Security Council in February 2006. Despite repeated calls by the U.N. Security Council for Iran to halt enrichment and reprocessing-related activities, and imposition of sanctions, Iran continues to develop enrichment capability at Natanz and at a site near Qom disclosed in September 2009. Iran insists on its inalienable right to develop the peaceful uses of nuclear energy, pursuant to Article IV of the NPT. Interpretations of this right have varied over time. Former IAEA Director General Mohamed ElBaradei did not dispute this inalienable right and, by and large, neither have U.S. government officials. However, the case of Iran raises perhaps the most critical question in this decade for strengthening the nuclear nonproliferation regime: How can access to sensitive fuel cycle activities (which could be used to produce fissile material for weapons) be circumscribed without further alienating non-nuclear weapon states in the NPT?

Leaders of the international nuclear nonproliferation regime have suggested ways of reining in the diffusion of such inherently dual-use technology, primarily through the creation of incentives not to enrich uranium or reprocess spent fuel. The international community is in the process of evaluating those proposals and may decide upon a mix of approaches. At the same time, there is debate on how to improve the IAEA safeguards system and its means of detecting diversion of nuclear material to a weapons program in the face of expanded nuclear power facilities worldwide.

(For more information, see CRS Report RL34234, *Managing the Nuclear Fuel Cycle: Policy Implications of Expanding Global Access to Nuclear Power*, coordinated by Mary Beth Nikitin; and CRS Report R41216, *2010 Non-Proliferation Treaty (NPT) Review Conference: Key Issues and Implications*, coordinated by Paul K. Kerr and Mary Beth Nikitin.)

Federal Funding for Nuclear Energy Programs

The following tables summarize current funding for DOE nuclear energy programs and NRC. The sources for the funding figures are Administration budget requests and committee reports on the Energy and Water Development Appropriations Acts, which fund DOE and NRC. The House passed its version of the FY2013 Energy and Water Development appropriations bill on June 6,

---

[120] The organization approved a new mission statement with the name change at its June 2010 meeting in Ghana. See http://www.gneppartnership.org.

2012 (H.R. 5325, H.Rept. 112-462). The Senate Appropriations Committee approved its version on April 26, 2012 (S. 2465, S.Rept. 112-164).

Table 2. Funding for the Nuclear Regulatory Commission
(budget authority in millions of current dollars)

	FY2010 Approp.	FY2011 Approp.	FY2012 Approp.	FY2013 Request	FY2013 House	FY2013 Sen. Comm.
Reactor Safety	806.8[a]	804.1[a]	800.1[a]	809.9	809.9	—[b]
Nuclear Materials and Waste	220.2	229.4	227.1	232.3	228.9	—
Yucca Mountain Licensing	29.0	10.0	0	0	10.0	—
Inspector General	10.9	10.1	10.9	11.0	11.0	11.9
Total NRC budget authority	1,066.9	1,052.3	1,038.1	1,053.2	1,059.8	1,054.1
—Offsetting fees	-912.2	-914.2	-909.5	-927.7	-921.7	-924.7
Net appropriation	**154.7**	**138.1**	**128.6**	**128.5**	**138.1**	**129.4**

a. Subcategories from NRC budget request.

b. Subcategories not specified.

Table 3. DOE Funding for Nuclear Activities (Selected Programs)
(budget authority in millions of current dollars)

	FY2010 Approp.	FY2011 Approp.	FY2012 Approp.	FY2013 Request	FY2013 House	FY2013 Senate
University programs	5.0	0	5.0	0	5.0	0
Reactor Concepts	—	168.5	115.5	73.7	126.7	73.7
Small Modular Reactor Licensing	—	—	67.0	65.0	114.0	65.0
Fuel Cycle R&D	136.0	107.6	187.4	175.4	138.7	193.2
Nuclear Energy Enabling Technologies	—	51.4	74.9	65.3	75.0	65.3
International Nuclear Energy Cooperation	—	3.0	3.0	3.0	3.0	3.0
Radiological Facilities Management	72.0	51.7	69.9	51.0	51.0	66.0
Idaho Facilities Management	173.0	183.6	155.0	152.0	162.0	152.0
Program Direction	73.0	86.3	91.0	90.0	90.0	92.0
Total, Nuclear Energy[b]	786.6	732.1	765.4	770.4	765.4	785.4
Civilian Nuclear Waste Disposal[c]	196.8	0	0	0	25.0	0

a. Not available.

b. Excludes funding provided under other accounts.

c. Funded by a 1-mill-per-kilowatt-hour fee on nuclear power, plus appropriations for defense waste disposal and homeland security.

Legislation in the 112th Congress

H.R. 301 (Forbes)

New Manhattan Project for Energy Independence. Establishes program to develop new energy-related technologies, including treatment of nuclear waste. Introduced January 18, 2011; referred to Committee on Science and Technology.

H.R. 617 (Matheson)

Radioactive Import Deterrence Act. Restricts imports of radioactive waste. Introduced February 10, 2011; referred to Committee on Energy and Commerce.

H.R. 909 (Nunes)

Roadmap for America's Energy Future. Includes provisions to triple the number of U.S. nuclear power plants, encourage recycling of spent nuclear fuel, develop nuclear waste disposal capacity, remove statutory limits on waste disposal at the proposed Yucca Mountain repository, establish a nuclear fuel supply reserve, and require NRC to establish expedited procedures for issuing new reactor combined construction and operating licenses. Introduced March 3, 2011; referred to multiple committees.

H.R. 1023 (Thornberry)

No More Excuses Energy Act of 2011. Includes provisions to prohibit NRC from considering nuclear waste storage when licensing new nuclear facilities, and to establish a tax credit for obtaining nuclear component manufacturing certification. Introduced March 10, 2011; referred to multiple committees.

H.R. 1242 (Markey)

Nuclear Power Plant Safety Act of 2011. Requires NRC to revise its regulation within 18 months to ensure that nuclear plants could handle major disruptive events, a loss of off-site power for 14 days, and the loss of diesel generators for 72 hours. Spent fuel would have to be moved from pool to dry-cask storage within a year after it had cooled sufficiently, and emergency planning would have to include multiple concurrent disasters. NRC could not issue new licenses or permits until the revised regulations were in place. Introduced March 29, 2011; referred to Committee on Energy and Commerce.

H.R. 1268 (Lowey)

Nuclear Power Licensing Reform Act of 2011. Requires evacuation planning within 50 miles of U.S. nuclear power plants and that reactor license renewals be subject to the same standards that would apply to new reactors. Introduced April 7, 2011; referred the Committee on Energy and Commerce.

H.R. 1280 (Ros-Lehtinen)/S. 109 (Ensign)

Requires congressional approval of agreements for peaceful nuclear cooperation with foreign countries. House bill introduced March 31, 2011; referred to Committee on Foreign Affairs. Senate bill introduced January 25, 2011; referred to Committee on Foreign Relations.

H.R. 1320 (Berman)

Nuclear Nonproliferation and Cooperation Act of 2011. Requires additional nonproliferation conditions for new peaceful nuclear cooperation agreements. Introduced April 1, 2011; referred to Committee on Foreign Affairs.

H.R. 1326 (Fortenberry)/S. 640 (Akaka)

Furthering International Nuclear Safety Act of 2011. Requires U.S. delegation to the Convention on Nuclear Safety to encourage member countries to use metrics in assessing safety improvements and publicly post national safety reports, and that U.S. agencies submit a strategic plan for international nuclear safety cooperation. Senate bill introduced March 17, 2011; referred to Committee on Foreign Relations. House bill introduced April 1, 2011; referred to Committee on Foreign Affairs.

H.R. 1436 (Christopher H. Smith)

Requires nuclear power facilities to notify NRC and state and local governments within 24 hours of an unplanned release of radionuclides above allowable limits. Introduced April 7, 2011; referred to Committee on Energy and Commerce.

H.R. 1694 (Engel)

Nuclear Disaster Preparedness Act. Requires the President to issue guidance for federal response to nuclear disasters, covering specific topics listed in the bill. Introduced May 3, 2011; referred to Committee on Transportation and Infrastructure.

H.R. 1710 (Burgess)

Nuclear Used Fuel Prize Act of 2011. Authorizes the Secretary of Energy to establish monetary prizes for advancements in used nuclear fuel management technology. Introduced May 4, 2011; referred to Committees on Science, Space, and Technology and Ways and Means.

H.R. 2075 (Engel)

Dry Cask Storage Act. Requires spent nuclear fuel to be moved from storage pools to dry casks within one year after it has sufficiently cooled. Owners of spent fuel could reduce their payments to the Nuclear Waste Fund to offset extra dry cask storage costs resulting from the act. Introduced June 1, 2011; referred to Committee on Energy and Commerce.

H.R. 2133 (Matheson)/S. 1220 (Conrad)

Fulfilling U.S. Energy Leadership (FUEL) Act. Among other provisions, authorizes nuclear fuel cycle research and development, including waste treatment processes and advanced waste forms. Requires the Secretary of Energy to consider recommendations of the Blue Ribbon Commission on America's Nuclear Future in implementing the authorized program and to submit a report to Congress comparing the Secretary's proposed long-term nuclear waste management solutions with the proposed Yucca Mountain repository. House bill introduced June 3, 2011; referred to multiple committees. Senate bill introduced June 16, 2011; referred to Committee on Finance.

H.R. 2354 (Frelinghuysen)

Energy and Water Development Appropriations for FY2012. Provides funding for NRC and DOE nuclear energy programs. Introduced and reported as an original measure by the House Appropriations Committee June 24, 2011 (H.Rept. 112-118). Passed House July 15, 2011, by vote of 219-196. Reported by Senate Appropriations Committee September 7, 2011 (S.Rept. 112-75). Considered on Senate floor November 16, 2011. Enacted as part of Consolidated Appropriations Act for FY2012 (P.L. 112-74), December 23, 2011.

H.R. 2367 (Pearce)

Government Waste Isolation Pilot Plant Extension Act of 2011. Would authorize disposal of government-owned non-defense transuranic waste in the Waste Isolation Pilot Plant (WIPP), in addition to currently authorized defense waste. Introduced June 24, 2011, referred to Committees on Energy and Commerce and Armed Services.

H.R. 3302 (Rooney)

Restore America Act of 2011. Among other provisions, would encourage tripling of U.S. nuclear power capacity, require licensing proceedings to continue for the proposed Yucca Mountain waste repository, remove statutory capacity limits on the repository, prohibit the President from blocking or hindering nuclear spent fuel recycling, establish a nuclear fuel reserve, and establish expedited reactor licensing procedures. Introduced November 1, 2011; referred to multiple committees.

H.R. 3308 (Pompeo)/S. 2064 (DeMint)

Energy Freedom and Economic Prosperity Act. Among other provisions, would terminate production tax credit for electricity generated by advanced nuclear plants. House bill introduced November 2, 2011; referred to Committees on Ways and Means and Energy and Commerce. Senate bill introduced February 6, 2012; placed on the Senate legislative calendar.

H.R. 3657 (Terry)

Nuclear Emergency Re-establishment of Obligations Act. Establishes criteria and procedures for the exercise of emergency authority by the NRC Chairman. Introduced December 13, 2011; referred to Committee on Energy and Commerce.

H.R. 3822 (Lowey)

Requires NRC to distribute safety-related fines collected from nuclear facilities to the counties in which the facilities are located to maintain radiological emergency preparedness plans. Introduced January 24, 2012; referred to Committee on Energy and Commerce.

H.R. 5325 (Frelinghuysen)/S. 2465 (Feinstein)

Energy and Water Development and Related Agencies Appropriations Act, 2013. Includes funding for DOE nuclear energy programs and NRC. House bill introduced and reported as an original measure by the Appropriations Committee on May 2, 2012, and passed House June 6, 2012, by vote of 255-165 (H.Rept. 112-462). Senate bill introduced and reported as an original measure by the Appropriations Committee April 26, 2012 (S.Rept. 112-164).

H.R. 4301 (Duncan)

Includes a requirement that NRC reach a determination on DOE's license application for the Yucca Mountain repository and removes existing statutory limits on the amount of waste that can be placed into the repository. Introduced March 29, 2012; referred to multiple committees.

H.R. 4625 (Joe Wilson)/S. 2176 (Graham)

Yucca Utilization to Control Contamination Act/Nuclear Waste Fund Relief and Rebate Act. Requires that payments into the Nuclear Waste Fund be returned to utilities unless the President certifies that Yucca Mountain is the selected site for a nuclear waste repository; that defense nuclear waste be transported to Yucca Mountain beginning in 2017; and that statutory requirements for disposal of nuclear waste be sufficient grounds for NRC to determine that waste from new or relicensed reactors will be disposed of in a timely manner. House bill introduced April 25, 2012; referred to Committee on Energy and Commerce. Senate bill introduced March 8, 2012; referred to Committee on Energy and Natural Resources.

S. 512 (Bingaman)

Nuclear Power 2021 Act. Authorizes a cost-shared program between DOE and the nuclear industry to develop and license standard designs by 2021 for two reactors below 300 megawatts of electric generating capacity, including at least one no larger than 50 megawatts. Introduced March 8, 2011; referred to Committee on Energy and Natural Resources.

S. 1320 (Murkowski)

Nuclear Fuel Storage Improvement Act of 2011. Authorizes the Secretary of Energy to provide payments to units of local government that, with the approval of the state governor, volunteer to host a "privately owned and operated temporary used fuel storage facility." Introduced June 30, 2011; referred to Committee on Environment and Public Works.

S. 1394 (Webb)

Allows a Commissioner of the Nuclear Regulatory Commission to continue to serve on the Commission if a successor is not appointed and confirmed in a timely manner. Introduced July 20, 2011; referred to Committee on Environment and Public Works.

S. 1510 (Bingaman)

Clean Energy Financing Act of 2011. Establishes Clean Energy Deployment Administration to provide financial assistance to commercial projects using clean energy technology, including nuclear power. Introduced and reported as an original measure by the Committee on Energy and Natural Resources August 30, 2011 (S.Rept. 112-47).

S. 2031 (Sherrod Brown)

Authorizes $150 million to demonstrate USEC centrifuge technology. Introduced December 17, 2011; referred to Committee on Energy and Natural Resources.

S. 2146 (Bingaman)

Clean Energy Standard Act of 2012. Establishes minimum U.S. annual percentages of clean energy use, including nuclear power, starting at 24% in 2015 and rising to 84% in 2035. Introduced March 1, 2012; referred to Committee on Energy and Natural Resources. Committee hearing held May 17, 2012.

Author Contact Information

Mark Holt
Specialist in Energy Policy
mholt@crs.loc.gov, 7-1704